Lost Geographies of Power

John Allen

Blackwell
Publishing

© 2003 by John Allen

350 Main Street, Malden, MA 02148-5018, USA
108 Cowley Road, Oxford OX4 1JF, UK
550 Swanston Street, Carlton South, Melbourne, Victoria 3053, Australia
Kurfürstendamm 57, 10707 Berlin, Germany

The right of John Allen to be identified as the Author of this Work has been asserted in accordance with the UK Copyright, Designs, and Patents Act 1988.

All rights reserved. No part of this publication may be reproduced, stored in a retrieval system, or transmitted, in any form or by any means, electronic, mechanical, photocopying, recording or otherwise, except as permitted by the UK Copyright, Designs, and Patents Act 1988, without the prior permission of the publisher.

First published 2003
by Blackwell Publishers Ltd, a Blackwell Publishing company

Library of Congress Cataloging-in-Publication Data

Allen, John, 1951–
Lost geographies of power / John Allen.
   p. cm. – (RGS-IBG book series)
Includes bibliographical references and index.
   ISBN 0-631-20728-7 (alk. paper) – ISBN 0-631-20729-5 (pbk. : alk. paper)
   1. Human geography. 2. Power (Social sciences) I. Title. II. Series.
   GF50 .A453 2003
   303.3–dc21   2002071211

A catalogue record for this title is available from the British Library.

Set in 10/12 pt Plantin
by Kolam Information Services Pvt. Ltd, Pondicherry, India
Printed and bound in the United Kingdom
by MPG Books Ltd, Bodmin, Cornwall

For further information on
Blackwell Publishing, visit our website:
http://www.blackwellpublishing.com

Lost Geographies of Power

RGS-IBG Book Series

The *Royal Geographical Society (with the Institute of British Geographers) Book Series* provides a forum for scholarly monographs and edited collections of academic papers at the leading edge of research in human and physical geography. The volumes are intended to make significant contributions to the field in which they lie, and to be written in a manner accessible to the wider community of academic geographers. Some volumes will disseminate current geographical research reported at conferences or sessions convened by Research Groups of the Society. Some will be edited or authored by scholars from beyond the UK. All are designed to have an international readership and to both reflect and stimulate the best current research within geography.

The books will stand out in terms of:

- the quality of research
- their contribution to their research field
- their likelihood to stimulate other research
- being scholarly but accessible.

Published

Geomorphological Processes and Landscape Change: Britain in the Last 1000 Years
David L. Higgitt and E. Mark Lee

Globalizing South China
Carolyn Cartier

Lost Geographies of Power
John Allen

Forthcoming

Geographies of British Modernity
David Gilbert, David Matless and Brian Short

Domicile and Diaspora
Alison Blunt

An Historical Geography of Science Outdoors
Simon Naylor

Contents

	Series Editors' Preface	vii
	Acknowledgements	viii
	Chapter 1 Introduction: Lost Geographies	1
Part I	**Spatial Vocabularies of Power**	13
	Chapter 2 Power in Things: Weber's Footnotes from the Centre	15
	Chapter 3 Power through Mobilization: From Mann's Networked Productions to Castells's Networked Fictions	38
	Chapter 4 Power as an Immanent Affair: Foucault and Deleuze's Topological Detail	65
Part II	**Lost Geographies**	93
	Chapter 5 Power in its Various Guises (and Disguises)	95
	Chapter 6 Proximity and Reach: Were there Powers at a Distance before Latour?	129
	Chapter 7 Placing Power, or the Mischief Done by Thinking that Domination is Everywhere	159

Chapter 8 Conclusion: Misplaced Power	189
Bibliography	198
Index	209

Series Editors' Preface

The RGS/IBG Book series publishes the highest quality of research and scholarship across the broad disciplinary spectrum of geography. Addressing the vibrant agenda of theoretical debates and issues that characterize the contemporary discipline, contributions will provide a synthesis of research, teaching, theory and practice that both reflects and stimulates cutting-edge research. The series seeks to engage an international readership through the provision of scholarly, vivid and accessible texts.

Nick Henry and Jon Sadler
RGS-IBG Book Series Editors

Acknowledgements

Books, especially ones with geography in the title, are the product of many connections, collaborations and friendships. If I said that this book was a collective venture, I would not be far wrong in terms of the conversations, ideas, provocations and flights of fancy that have crossed my path in the course of writing it.

As a head of department at the Open University for the best part of the last six years, I can safely say that I have learnt little about the nature of power from that experience and so have had to look elsewhere for inspiration and insight. I found that through the ample comments on earlier drafts of the chapter materials, invitations to write, and seminar presentations at various university departments, and so I wish to extend my thanks to all those who have helped me with my efforts: John Agnew, Ash Amin, John Clarke, Allan Cochrane, Chris Hamnett, Nick Henry, Steve Hinchliffe, Alan Hudson, Roger Lee, Andrew Leyshon, Doreen Massey, Joe Painter, Chris Philo, Steve Pile, Michael Pryke, Jenny Robinson, Andrew Sayer, David Slater, Kristian Stokke, Grahame Thompson, Nigel Thrift and Jane Wills.

I have also gained much over the years from just being a part of the Geography Department at the Open University, especially from the many exchanges with colleagues and research students – with their obsessions and wide-ranging interests – and not least from their companionship and humour (often it should be said in response to my ideas).

On a more practical level, thanks are also due to Sarah Falkus and Angela Cohen at Blackwell Publishing for their encouragement and advice in preparing this book for publication. I am also in debt to Jan Smith and Sylvia Laverty at the Open University for their professional secretarial support, the value of which is inestimable in a climate where such support has all but disappeared from academic circles.

Finally, thanks go to Jo Foord for her generosity with both ideas and time, as well as the encouragement she has given me when I flagged. That just leaves me to acknowledge my relationship with Jack and Adam, my two sons, who by being themselves gave me something beyond power: a sense of what matters.

John Allen

1

Introduction: Lost Geographies

It might seem odd to suggest that geography is something that we can lose. It may seem even odder to suggest this can happen around power and its relationships, especially as geography and power seem to run together in so many ways. The connection of geography with power, if one thinks about it, is pretty much a familiar one. Most political disputes over land and territory, in Europe and beyond, where borders have been torn up and redrawn by coercive states or countries subjected to the dominant force of neighbouring governments or ethnic groupings, have geography at their core. Closer to home, the gated communities which have sprung up in major cities to enable the affluent to live behind high walls and electronic gates are an integral mix of geographical and economic constraint. Then there are those unsettling moments when you find yourself on the receiving end of a blunt decision or insensitive instruction taken by some far-off government agency or impersonal corporation and can only begin to wonder where such powers at a distance really come from, let alone who lies behind them. Or the times, in a public space perhaps, where you are made to feel that your every movement is under observation, subject to surveillance of some kind, yet are quite unable to say whether this is actually happening or that anyone is really directing things.

Yet the argument of this book is that however familiar the association of geography with power, we have lost the sense in which geography makes a difference to the exercise of power. For all that someone like the French philosopher Michel Foucault might have told us about power turning up more or less everywhere because it comes from everywhere, the landscapes of power that are familiar to us have, to my mind, limited rather than extended our understanding of power. In a world where it has almost become commonplace to talk about power as networked or concentrated, distributed or centralized, even decentred, deterritorialized or radically dispersed, it is all too easy to miss the diverse geographies of power that put us in place.

As I see it, power is not something that is simply extended over short or long distances, or something which radiates out from an identifiable central point, or something which engulfs places in ways that are all pervasive. Power is not some 'thing' that moves and it does not traverse and transect places or communities, so that we may be forgiven for thinking that it is all encompassing. Power, as I understand it, is a relational effect of social interaction. It may bridge the gap between here and there, but only through a succession of mediated relations or through the establishment of a simultaneous presence. People are placed by power, but they experience it at first hand through the rhythms and relationships of particular places, not as some pre-packaged force from afar and not as a ubiquitous presence. The diverse geographies of power that I wish to foreground work through relations of proximity and reach to bring the far-off into people's lives and also to make the close-at-hand sometimes feel remote.

It is this grasp of what geography can do that has been lost, or rather perhaps I should say it is this particular geography of power that is waiting to be understood for the first time.

Much of this book is given over to reminding us of the difference that geography can make to the exercise of power: that power in its various guises takes effect through distinctive relations of proximity and reach, and in so doing exercises our lives in ways that are not always so familiar. Only by turning over some of our familiar assumptions about geography and power can we glimpse some of the many ways in which power puts us in place. It is in this sense that geography matters to our relationships with those who exercise power and, turning this around, what power is matters to what can be said geographically about its exercise.

For my part, I wish to press this claim through two lines of argument. The first is that in the rush to see power as something which turns up more or less everywhere, I think that we have lost sight of the *particularities* of power, the diverse and specific *modalities* of power that make a difference to how we are put in our place, how we experience power.

Believing with Max Weber and Hannah Arendt that power is never power in general, but always power of a particular kind, I take such acts as domination, authority, seduction, manipulation, coercion and the like to possess their own relational peculiarities. The glib sense in which power is believed to be all around us has not only proved to be a source of distraction in this respect, it has also to my mind minimized the experience of what it means to have had a brush with power. A world of difference separates dominant relationships which restrict choice and close down possibilities from those which, for instance, secure assent, manipulate outcomes, impute threats or seduce through suggestion and enticement. I want to bring such differences and distinctions back to the fore, not only to underline the kinds of confrontation that power can and does take, but also to weave them into a more geographically curious dialogue of power.

For the second part of my argument is that I think that we have lost the sense in which power is *inherently* spatial and, conversely, spatiality is *imbued* with power. Much of what I take to be the diverse geographies of power has been lost to our understanding precisely because the likes of domination, authority and seduction have not been thought through in terms of how they are exercised, whether close at hand, at a distance, through a succession of relationships or established simultaneously.

Authority, for instance, is a peculiarly distinctive act, as anyone who has ever been mocked or undermined whilst trying to exercise it will readily testify. Authority, whether it be the discipline laid down by a teacher early on in life, or later by a managerial figure of some kind, works through recognition in an attempt to secure the willingness of others to comply. Such authority can be geographically far-reaching or it can be exercised close at hand, but the more distant it is, the greater the possibility for both the position and the act to pass unrecognized. Presence and proximity matter to authority in a way that to mock or to poke fun at, say, a dominant force like a multinational firm with a monopolistic grip on the market does not. It does not take long to realize that lives subjected to domination may quite easily be constrained and controlled either at arm's length or at close quarters. There is, it is important to stress, nothing predetermined in a spatial sense about any of this, nothing to suggest that authority will falter at a distance; it is merely that, given the type of relation that authority is, presence and proximity will inevitably matter to its exercise.

Maybe the sense in which closeness matters to authority has only ever been acknowledged in a matter-of-fact way, but I do none the less wish to claim that we have lost sight of such spatial references, and in so doing impoverished our understanding of power. It seems that much easier to see the association between power and geography through the odd tall fence, high wall and exclusionary boundary marker than it is to recognize that the many and varied modalities of power are themselves constituted *differently* in space and time. Paradoxically, the more attention that has been paid to the spaces of power, its flow, circulation, diffusion, displacement even, the less we know, it seems to me, about how geography affects the workings of power. Somewhat ironically, we now have a richer spatial vocabulary of power than hitherto, but a poorer grasp of the difference that geography makes to the *exercise* of power. As such, a good deal more has been obscured than revealed.

This book tries to show how things might look if we stop treating power as uniform and continuous over space and start to think through the diverse geographies of power's proximity and reach, and how these play across one another. In this topological landscape, fixed distances and well-defined proximities fail to convey how the specific relational ties of power are established. For that, we need to be a little more curious about power's spatial constitution and not get lost or distracted in familiar spaces.

Before we adopt this geographically curious pose, I should first say a little more about what I take power to be.

Situating Power

Part of the familiar baggage of power can, I think, be traced to a particular view of it that is surprisingly hard to shake. Though many of us might like to distance ourselves from it, the sense in which power is out there ready to be wielded or clung on to or flaunted is deeply ingrained. It is almost inconceivable to imagine our lives separate from those bodies, institutions and figures who have this thing called power at their disposal. It is just part of the durable architecture of social life: some people have it but choose not to use it, whilst others use it and even abuse it.

We may dismiss such tangible images, preferring instead to think about power as something exercised rather than held, but that will not prevent such images from registering in our minds the next time some event or character upsets the balance of order. It is the figure of the slick chief executive who, in a bid to amass even greater wealth and fortune, inflates the company's size to buy influence and exploit trading loopholes; it is the bureaucrat who knows how to turn the rule-making machinery to their advantage and admit no discretion; or, more graphically, it is a modern-day battleship carrier group steaming over the horizon towards us at a rate of knots. Equivalent in height to something like a twenty-storey building, it is not difficult to see how such battleship formations could arouse a mixture of awe and fear among those looking on. Weber and Arendt may well have been right to stress that there is no one 'thing' called power, that the concept itself is 'amorphous', but that will not stop most of us from acting as if there were such an embodied entity.

Even if we cannot shrug off such thoughts and images, however, that does not mean that we have to believe in power of the sort that only moves people and mountains. To suggest that power is a *relational effect*, that it is an outcome of social interaction not something designed to put a blunt stop to it, does not convey quite the same convincing impression as a manipulative chief executive, but it is more in line with what we have been saying about power turning up more or less everywhere. If power is something that exercises us in particular ways, through various modalities, then we need to understand why it is misconceived to equate massed resources and capabilities with power.

Size and ability are perhaps among the more obvious reasons why we should think of resources as power. The bigger the capabilities at our disposal, whether measured by financial muscle, skills, information, contacts or sheer fire-power, the greater the assumed power. Why this should be so is not entirely clear to me, as indeed it was not to Anthony Giddens, or to the American sociologist Talcott Parsons before him or, for that matter, the able Florentine Niccolò Machiavelli back in the sixteenth century.

Resources may be misused, incompetently applied, mobilized for all the wrong reasons and, perhaps worst of all, simply wasted by a misguided yet otherwise well-meaning bunch of individuals. Those in charge may make a string of bad decisions, or those nominally in control may pool all available resources, yet to no avail. What can go wrong may well go wrong, and if it does not it may succeed only partially or hardly at all. In short, power as an outcome cannot and should not be 'read off' from a resource base, regardless of its size or scope. Power in this sense is no more to be found 'in' the apparatus of rule than sound is to be found 'in' the wood of musical instruments. It is, as suggested, a relational effect, not a property of someone or some 'thing'. Resources can increase in size and you can lose them or they may simply evaporate, but I am not convinced that power can do any of these things. Power, it seems to me, is often disguised as resources and in that sense we need to disentangle the two; we need to distinguish clearly between the exercise of power and the resource capabilities mobilized to sustain that exercise. I draw this distinction as the basis for a more thorough examination of the specific relational ties through which power establishes itself.

These ties, broadly speaking, take one of two forms: either *instrumental*, where power is something that is held over you and used to obtain leverage, or *associational*, where power acts more like a collective medium enabling things to get done or facilitate some common aim. The contrast is significant because it is what we experience when power is exercised either *over* or *with* others. Whereas the former is always exercised at someone else's expense, excluding them or putting them in their place in ways that constrain, the latter represents a means of enablement where all those taking part may benefit in some way. If we think of the former as the archetypal view of power, an action rooted in conflict where one side subjects the other to its will, the latter, unaccustomed view of power is rooted in mutual action and holds out the prospect of empowerment for all those involved. As a convenient shorthand, one involves the exercise of 'power over' others and the other the 'power to' act.

The shorthand is not mine alone; it has a long and varied history as it has been adapted to the concerns and interests of diverse scholars (see Allen 2002). I use it as a guide more than anything else to return our attention to the *particularities* of power, those relationships around us that we experience as constraint, which sometimes touch us in ways that are often hard to pinpoint, and those we experience as enabling through various forms of association.

Authority, as I have said, is a peculiarly distinctive act which works through recognition. Once claimed, whether by doctors, teachers and lawyers or just plain ordinary managers, it has to justify itself in the eyes of those around them. It is conceded, not imposed, and even to those in professional office, where the position goes with the job so to speak, authority is 'lent', and only for so long as recognition lasts. Compliance is always conditional and anyone thinking that a rule book is all the legitimacy that is necessary is one day likely to be in for a rude awakening. I stress this distinctiveness not because I am vexed by deep semantic anxiety, but to avoid the easy reaction to talk of power being everywhere, which is to slide between power, authority and domination *as if* they are the same thing in terms of their effect.

Authority, as an instrumental act, is exercised like *domination* at someone else's expense, but it does not involve the imposition of a form of conduct so that submission is the only possible option. Nor does authority involve the *manipulative* concealment of intent, or the *coercive* threat of force to exact compliance, or the *seductive* arousal of this rather than that line of interests. It is these differences that I want to bring back to the forefront of discussion, as well as those between *negotiation* and *persuasion*, for example, which can speak to a less confrontational agenda of power.

Inevitably, there will always be some slippage of meaning as far as power is concerned, but that does not take anything away from the fact that different modalities of power work through distinct relational ties. Omnipotent chief executives and bureaucratic monoliths, as well as any number of 'exclusionary' arrangements, will no doubt continue to carry the day in terms of the durable architecture of power, but the powers all around us are powers of a particular kind – with quite specific effects. And, unlike so-called blocs of power, they are inextricably spatial.

From Spatial Vocabularies of Power...

When we turn to existing vocabularies of space and power, the diverse geographies of power that I have in mind elude our grasp more easily, if anything. As previously mentioned, a major part of the argument of this book is that our new-found appreciation of all matters spatial has served to obscure rather than reveal much of the difference that geography makes to the exercise of power. In part I, I devote considerable attention to the spatial vocabularies of power employed by different writers, with an eye to the explicit spatial characteristics that each employs within their broad frameworks of meaning.

The astute reader will notice the absence of certain key writers on power, most notably the political sociologist Steven Lukes but also C. Wright Mills and the neo-Machiavellian elite theorists Vilfredo Pareto and Gaetano Mosca, and the reason for this is quite straightforward. Only those writers who possess more than a nominal sense of space in their writings on power have been included. It was never my intention to produce an A-to-Z of power from the political theorist Thomas Hobbes onwards. There are other texts available which admirably provide such coverage. The coverage I seek, in contrast, ranges from those writers with little to those with a rather fertile *spatial imagination* of power.

At the barren end of the spectrum, Max Weber, for instance, talks about the powers of command and their distribution, although where space appears as little more than a complication to its delegation or distribution. On a more engaged basis, Anthony Giddens speaks about distanciated forms of power, where relationships stretched over space represent a more modern, facilitative means of securing outcomes, and Michael Mann blends authoritative and diffuse techniques of organization to show how overlapping networks of power can achieve far-flung goals of one kind or another. Manuel Castells, too, provides an engaging account of how power flows through complex systems of networked interaction.

No assessment of power relations, space and spatiality would be credible, however, unless it included the work of Michel Foucault and Gilles Deleuze. Foucault's attention to constitutive detail, where the arrangement of spaces which make up institutional complexes, like the prison or the clinic, are seen as integral to the ways in which particular forms of conduct are secured, lies behind much of the recent fascination with everyday spaces of power. His interest in the art of dispersed government too, where entire populations are said to come under the sway of a broad range of immanent techniques oriented towards the production of new subjectivities, has also proved to be influential. And Deleuze, with his co-writer Felix Guattari, has written about deterritorialized apparatus of rule in ways that have attempted to convey the pervasive nature of contemporary power. Indeed, a diagrammatic sensibility, where power is seen to constitute its own organization yet produce itself relationally from point to point, is common to the spatial vocabularies of Foucault and Deleuze.

For all such writers, space in one way or another is implicated in relationships of power; it is significant to their realization or actualization. Some vocabularies are a little more stilted than others in the way that they represent spatiality and some more challenged than challenging in their view of space, but there is none the less a spatial imagination to reckon with, especially in the writings of Foucault and Deleuze. Yet for all their labours, I think that, for different reasons, they fall short of an appreciation of the inherent spatiality of power and, relatedly, that spatiality is itself imbued with power.

Such a principal lack, in my view, is evident in the case of Weber and indeed wherever power is represented as a 'thing-like' property capable of extensive reach. The possession of power identifies its location – it is in the hands of someone, marshalled by somebody, concentrated in some institution – and in a comforting sort of way its extension or distribution over space is taken for granted. There is a kind of homely promise that nothing much really happens between here and there to cause us to worry about what moves exactly or what, if anything, is distributed. Even in the more interesting contemporary landscape of a territorially reordered world where political governance has to negotiate a more complex institutional geography – between, say, global institutions like the International Monetary Fund (IMF) and the World Bank, footloose multinationals and the now ubiquitous non-governmental organizations (NGOs) – power still seems to be regarded as something which shifts across borders or is redistributed between sites of authority without too much difficulty. It may well be that there has been some redistribution of rights and responsibilities between global actors, but it strikes me that for much of the time what passes for power is some pre-formed capacity that is more or less transmitted intact across the landscape. Geography, it would appear, is little more than a minor disruption to the general exercise of power.

In the systematizing hands of Mann and Castells there is rather more to space and spatiality than this, and yet they too, in my view, rely upon an equally unproblematic account of how power 'travels' or flows through the multiple networks that comprise society. True, they both produce rich accounts of networked interactions in their writings, where power (re)-sources are organized in a loose or tightly orchestrated fashion, but it is hard to get away from the fact that power, for them, is something that is generated at different sites across the networks and projected with relative ease between locations. Because power is considered to be a 'fluid' medium, it is almost as if they wish to suggest an analogy with the circuitry of electrical energy, where networks transmit power as a matter of course through any number of connected circuits. Indeed, perhaps because a view of social space which foregrounds the circulation and flow of power is only too recognizable, it appears almost pointless to question it.

But if, as argued, power is not some 'thing' or attribute that can be possessed, I do not believe either that it can flow: it is only ever mediated as a relational effect of social interaction. Put another way, power is not a uniform or continuous substance transmitted across tracts of space and time; it is always constituted in space and time.

This is the kind of appreciation that both Foucault and Deleuze bring to an understanding of power and spatiality. Subjects are constituted by the spacing and timing of their own practices as much as they are by those who seek to shape their conduct. As an immanent force which constitutes its own organization, not one imposed from above or from the outside, power is seen as coextensive with its field of operation. Power is practised before it is possessed and it is this that gives rise to the roundaboutness of power, not some facile notion that it is a shadowy force lurking in the murky recesses. The spaces of the everyday are the sites through which subjectivity is immanently produced.

Or, rather, the commonplace institutional spaces of education, health, welfare, work and correction provide the settings through which power is exercised. But how such arrangements work outside of the walls of the institution, in the context of a widely diffuse and disparate population, seems to me to demand an altogether different spatial imagination. If the art of dispersed government is to mean anything at all, I would argue that we need to go beyond the spread of 'certainties' outwards to consider the more mediated relations through which the far-off is brought within reach. More than that, we need to be a little more curious about power's spatial constitution in a landscape that does not assume fixed distances, well-defined proximities and effortless reach.

...to Power's Spatial Constitution

In part II, I hope to be able to show what we have lost that is geographical in relation to power. For instance, I consider Weber's grasp of the subtle distinctions and modalities of power to be insightful, yet what is gained in analytical strength is lost by his less than convincing view that power is simply impelled outwards. If he had problematized the spatial in the way Foucault endeavoured to do, then a more coextensive notion of power's twists and turns might have been possible. Equally, if Foucault had reflected a little more on the difference between domination and authority, and how they in turn differ from the practice of manipulation or the act of inducement or the art of seduction, then a more nuanced account of spatiality and power might have been attainable.

But this is merely wishful thinking. In part II, I somewhat ambitiously aim to blend the two lines of analysis – to show how spatiality is constitutive of power relations not only in general, but also in the particular ways in which, say, seduction takes advantage of existing attitudes and expectations to reach an audience or domination works to constrain a disparate population. Without a grasp of the particularities of power we cannot begin to understand the difference that geography makes to its exercise. Even if we accept Foucault's immanent conception of power, as I broadly do, it is important to bear in mind – lest we lose sight of it – that power relations have long been experienced through a variety of different modes and that they are always already spatial.

As I see it, there is no spatial template for power, but that does not mean that power assumes some kind of undifferentiated spatiality. Authority, to press the point, works through relations of proximity and presence if it is to be at all effective in drawing people into line on a day-to-day basis. The more direct the presence, the more intense the impact. The same goes for coercion, that most certain imprint of power, where the threat of force holds only for as long as people feel restrained by its possibility. Manipulation, in contrast, is a one-sided affair where the concealment of interests gives its application significant spatial reach, as do the modest, suggestive qualities of seduction, which leaves open the possibility for people to reject or remain indifferent to its pervasive exercise. For seduction is a hit-or-miss type of power, where the possibility of refusal is built in, so to speak. Domination as an act of blanket constraint is different again in terms of both its social and spatial constitution, as is the reward-based nature of inducement, and so on and so forth.

To belabour the point, though, the issue is not one of detailing the geographical aspects of power, but, as I have said, one of grasping the *particular* ways in which modes of power take effect. There are no pre-formed strategies of domination back at base waiting to be unleashed or off-the-peg seductive acts in the catalogue available on request. There are only power's mediated relations, which may draw the more or less dispersed lives of people closer through real-time technologies or reach out to them through a succession of relations and practices. Either way, the relationship between proximity and presence plays across the gap between here and there in different ways depending upon the *specific* ties of power in question. It is in this sense that I suggest geography makes a difference to what we experience as power and to how it is exercised.

Or rather, I should say that this is only the half of it.

When you are placed *within* a tangled arrangement of power relationships, the all too familiar spaces of the office, the workplace, the estate or the community, even the 'breathing spaces' we thought were our own may look and feel quite different. Thinking about how power reaches into our lives gives us a different vantage point from being immersed in its cross-cutting arrangements. In this topological landscape, the close-at-hand can be made to feel just as distant as the remote elsewhere, as people move in and around one another closing down possibilities, smothering choices and broadly making us feel that we do not belong – as if we were part of someone else's space. This is the kind of dominant presence that Henri Lefebvre frequently portrayed in his work, but there is, I think, more going on in the name of power in such places than certainly he would have credited.

In attempting to unravel the rhythmic and routine complexity of familiar spaces, however, Lefebvre did provide another angle on what it means to talk about the roundaboutness of power. Through the constant succession of movement and activities, the manner in which they are performed and the style in which they are executed, places take on a life of their own, with certain groups able to superimpose their presence on others. In the entangled nature of people's lives, places, on this account, take their shape through dominant or controlling rhythms that seek to suppress the routine traces of others. Exclusion in this context has less to do with closed doors and high walls, and rather more to do with spaces constructed by dominant groups in their own likeness – through a series of rituals and gestures, moods and attachments, as well as accumulated styles and meanings. The composition of space, the partition and layout of particular uses, also serves as both a resource and the means through which power is exercised.

There are resonances with Foucault's diagrammatic sensibilities here, perhaps more than Lefebvre was prepared to acknowledge, but what characterizes such power-laden spatial circumstances, in my view, is the diversity of power relations which exercise us. If we were to probe many a familiar space, institutional or otherwise, domination is unlikely to be the only powerful presence. Spaces may be laid out for temptation in a seductive way through a combination of suggestive practices, inclusive designs and enticing layouts, or they may be subject to manipulation by groups constructing them in their own likeness, which conceal or disguise their true motives. Ritualized ways of doing things, say in the familiar setting of the school, the clinic, the bank, the law courts, the church or the mosque, may be conducted in a manner that lends weight to the figures who work there and give them authority. The buildings themselves, in the case of the law court for instance, may serve as a symbol of authoritative values 'demanding' recognition and respect. Once inside its walls, the rhythm of activities, the lulls and outbursts, may generate their own coded style of authority. And so on.

For my part, it is not that I wish to convince you that all places are saturated with the fixture and fittings of power. On the contrary, particular places may play host to a variety of cross-cutting arrangements of power, or such traces may be strikingly absent. But if power has a presence at all, it has it through the interplay of forces established *in place*. People are placed by power, but not as the result of some massed force transmitted intact by some central administration from up the road or even from the other side of the globe. The arrangements of power we find ourselves exercised by may well arise from ideas and events hatched elsewhere, but that, as I understand it, is merely another way of saying that the presence of power is more or less *mediated* in space and time.

In fact the most challenging aspect of all this is to try to understand the diverse geographies of power that seek to put any number of us in place; that is, the different ways in which relations of proximity and presence play across the gap between here and there to bring the far-off within reach, yet for others make the nearby seem closed off and distant. Whether on the receiving end of some biotechnology giant intent on introducing genetically modified materials into the food chain, or subject to the controlling rhythms of an institutional force which makes you feel 'out of place' in your own familiar surroundings, or seduced into thinking that the politics of privatized welfare has much to offer, it matters that we understand *how* power exercises us. For it is through this varied understanding, where we are able to recognize the powerful and not so powerful forces that face us, that the possibility for *empowerment* lies. It is in this sense that a less familiar understanding of geography and power can hold out the prospect of alternative, more collaborative, paths to action and social change.

Part I

Spatial Vocabularies of Power

In the first part of the book, the term 'spatial *vocabulary* of power' is employed to convey the fact that in many recent studies of power and power relations, space is indeed considered important to its exercise. Space has a role to play in the overall description of how power manifests itself or how it is realized in different contexts and settings. But, as I argue in each of the following chapters, the understanding of spatiality outlined never really gets beyond a richer vocabulary in which power is either distributed or extended over space, or regarded as something which flows or has the ability to be all around us in some kind of immanent fashion. As a result, we are offered a documentation of the spatial aspects of power, rather than an examination of what it means to think through the diverse effects of power spatially.

If this sounds rather precious around the issue of space and spatiality, it is not my intention to dwell on the minutiae of differences between various authors' understanding of the latter. Rather I hope to be able to show how going beyond an understanding of the scalar landscape as one of fixed distances, well-defined proximities and uncomplicated reach has much to offer, not least for what it means to experience a brush with power.

2

Power in Things: Weber's Footnotes from the Centre

How often are we told that power is in the 'hands' of the professionals or some notable government figure or global institution, like the World Trade Organization? We may wish to quibble about the extent of their power or ask questions about who behind the scenes really has the power to set the agenda or fix solutions, but by and large it seems entirely reasonable to suggest that someone somewhere must have it. After all, it would seem a little odd to suggest that power has no reference point at all, no location to speak about.

The sense in which power is thought to have a location rests largely on the assumption that it is something that can be *held or possessed*, in much the same way as local political elites, managing directors and government ministers are said to possess the power to shape our lives. It then seems relatively straightforward to say that people and institutions 'hold' power and do so by drawing upon the range of abilities at their disposal. This view of power as tactile, as something that can almost be touched, is a compelling one. It is this quality which gives the impression that power may actually be located or stored in 'things' – in imperious or less authoritative institutions and the people who run them.

From this, it is but a short step to talking about powers held 'in reserve' or as a resource capability 'located' in the apparatus of institutions, be they public or private. On this account, power is considered as something which may be delegated or distributed, almost invariably from a centralized point to various authoritative locations across any given territory.

Bruno Latour (1986) has described this type of imagery as a 'centred' conception of power, where a central source is responsible for the diffusion of power within and across society. At its simplest, power is delegated or distributed in a relatively straightforward manner down through an organizational hierarchy under clear lines of authority, and there may be one of two outcomes. Either the rules, regulations and constraints imposed by the centre are successful in meeting its goals or their organizational impact is minimized or deflected by the degree of resistance met. As such, power is located in 'things' in a relatively transparent fashion, although its effects are perhaps not as easy to unravel or anticipate as Latour would have us believe. Indeed, whilst it may be possible to locate with a reasonable degree of accuracy which institutions and which individuals or groups 'hold' power, it remains uncertain as to how, when and where that capacity will be exercised.

Another way of expressing this is to draw attention to the *dispositional* quality of power that defines such capacities. If, for example, an institution has the capacity to influence and control the actions of others in a particular way, that capability is regarded as a latent ability, rather than an observable exercise of its powers. The capacity to do something is clearly distinct from the actual exercise of that capability. In other words, the power that the institution 'holds' is potential: it is known to be possessed by the institution, but it is only exercised under certain conditions. From the longstanding powers bestowed on professional groupings such as those practising medicine or law, to the prestige and symbolic power attributed to banking and other financial institutions today, or to the disciplinary powers inherent in the nineteenth-century agencies of colonial control that enabled them to wield authority at great distances, the capacity for domination is generally conceived as present and capable of being exercised should the need arise or circumstances dictate. Such bodies are said to have the power vested in them, regardless of whether or not they actually use that power. It is this possession of power – as a property, relation, or attribute – that reveals its 'centred' location.

Also revealed by the above is the ease with which the language of power as capacity translates into the language of power as domination. The idea that individuals or institutions possess specific capacities to secure the compliance of others can be thought about as just so many ways of achieving dominance over others. Such an instrumental view of power, whereby one's will is imposed on the actions of others, is strongly reminiscent of the line of thought that runs from Thomas Hobbes to Max Weber. On this instrumental view, in any situation of conflict or competition, those with the greater amount of power are often assumed to prevail over those with less. Indeed, writers like Barry Hindess (1996) take this quantitative conception of power as the cornerstone of any view of power which regards it as a general capacity. One side always gains at the expense of another, or so it would seem.

It is not necessary to adopt a quantitative conception of power, however, to adhere to a 'centred' conception of it, where the capacity to dominate is delegated or distributed over space to secure certain goals. In this chapter, attention is drawn to the – often implicit – spatial vocabulary of power that is entailed whenever power is represented as a capacity and to the limited understanding of the relationship between power and space that is involved. At worst, power is conceived as something which radiates out from an identifiable centre, triggering an effective capacity for control. Once you have the measure of power and a grasp of its capacity to administer, control and fix a territory and its population, the rest, it seems, amounts to little more than a series of footnotes from the centre. And even where power is no longer seen to operate in quite such a straightforward top-down or centre-out fashion, as for example in recent accounts of 'multi-level' or 'multi-tier' governance, I would argue that it rarely adds up to more than a reorganization of geographical scale, where power is more or less redistributed intact across the scalar landscape.

Before we consider these spatial issues, however, it is important to be clear about what is meant when power is referred to as a *capacity*.

Power as a Capacity

As mentioned above, the notion that power may take the form of a capacity which enables its holder to secure certain outcomes or realize certain objectives is a dispositional quality. In whatever way we may wish to think about it, it reduces to the proposition that power can be held without being exercised. In practice, however, the two aspects are often conflated, with the possession of power only readily acknowledged in its exercise.[1] It is not uncommon, for example, to 'read off' the power of the multinational 'giants' in the oil and petroleum industry on the basis of their past performance. The likes of Shell Oil, Texaco and Exxon Mobil, for instance, are assumed to be powerful on the basis of what they have done, or are alleged to have done, around the globe – dictate the terms of negotiation with host countries, play off local interests, dominate particular markets – regardless of whether or not they are presently engaged in such activities. It matters not one jot that the operational headquarters of such companies may fail to

---

1 In many ways, the classic political science debates on power in the 1960s and 1970s which ran from Robert Dahl (1957, 1961) to Steven Lukes (1974) and beyond revolved around Dahl's conflation of the possession and the exercise of power. For Dahl, the observable exercise of power – in the key political decisions taken – was all that really mattered, although he implicitly assumed that ruling groups possess power on some basis. In contrast, Lukes and others pointed to the latent interests and actions which influenced outcomes beyond the arena of observable conflict. The dispositional quality of power, however, was not a consideration as such in the debates, even though it was evident in outline. See Ted Benton (1981) for a critique of Lukes and projection of a dispositional view in terms of a realist philosophy.

recognize such 'leverage'; the point is they are deemed to possess it irrespective of where or indeed whether they use it. Their power is always already present, as a 'thing-like' quality capable of being marshalled and exercised. In this instance, power is likely to be attributed to such companies on the basis of their size and assumed performance.

With bureaucratic power the logic is likely to be different, in so far as power is deemed to be held on the basis of position within an organization rather than on any notion of institutional size. The managers in a hospital or their professional equivalent in a high-technology or oil company occupy their decision-making roles by virtue of their technical expertise and competence to do the job. Authority comes with the position and with it the structural limits of what an official may dictate or prescribe. There is nothing monolithic about the capabilities involved; in so far as power is exercised, this happens in a downward fashion within the strict limits of the rules laid down by the bureaucratic institution.

The capacity to act or dominate, however, is not restricted to institutions. Social groups may exercise their power in ways which serve to exclude or check the freedom of movement of others. The construction of gated communities in the inner city, for example, complete with high security walls and remote-controlled gates, can be understood as but the latest in a long line of instances of the middle classes flexing their economic capabilities to protect their lifestyles and differentiate themselves from the urban 'mass'. The capacity to exclude, to 'purify' a piece of inner-city space, is thus a possibility that the middle classes have at their disposal – should they choose to exercise it.

In each of the above contexts, however, the mere fact that corporate and bureaucratic institutions, along with the middle classes, share a set of capabilities tells us little about what these abilities comprise, or how institutions or people became endowed with them, or even that they refer to the same 'thing'. The notion of power as a capacity may, for example, refer to a set of causal properties intrinsic to the very nature of what it is to be a monolithic multinational corporation. Or it may refer to a range of interests attributed to affluent social groupings. Or indeed, the meaning of the term can be altogether looser, with the intention of conveying through metaphor something that is merely 'held' by social groups or institutions. As a means of disentangling the conceptual issues involved, it may be helpful to draw upon Barry Hindess's forthright assessment of power conceived as a capacity.

Scripted power plays?

Much of Hindess's *Discourses of Power* (1996) is given over to a critique of the view that power can be conceived in terms of a capacity which enhances those who possess it. Eschewing any notion of power-in-general in favour of a more heterogeneous conception based upon Foucault's techniques of self-regulation (discussed in chapter 4), Hindess identifies a number of shortcomings in the notion of power considered as a general capacity. Leaving to one side for the moment his metaphysical worry that there are those who appear to attribute a common substance to such a capacity, a kind of 'essence of effectiveness', Hindess's main concern is that the description of power as a capacity lends itself to a quantitative and deterministic conception of power:

> If to have power is always to possess the essence of effectiveness in some definite quantity, then those who possess a greater quantity of that essence obviously will be more effective than those with a lesser quantity... This view of power as a quantitative and mechanical phenomenon which determines the capacity of actors to realize their will or to secure their interests has been enormously influential in the modern period... The great attraction of this conception of power as quantitative capacity for so many social scientists is that it appears to promise an easy means of identifying who has power and who has not. (1996: 26–7)

Related closely to the idea that power is a quantitative capacity, therefore, is the assumption that it is possible to determine in advance who wins and who loses. If more power prevails over less, then the outcome of distributive struggles would appear to be simply a matter of the summation of the respective powers involved. Drawing upon Max Weber's definition of power to illustrate the asymmetrical nature of power relationships which flow from this assessment of quantitative power, Hindess concludes that power, on this view, may be used to command and control others:

> People employ power in their dealings with things and in their dealings with each other. In the latter case, this conception of power implies that the wishes of those with more power will normally prevail over the wishes of those with less. It is for this reason that Weber identifies power with 'the chance of a man or a number of men to realize their own will even against the resistance of others who are participating in the action' (Weber, 1978, p. 926). This conception of power as simple capacity suggests that there will be an unequal relation between those who employ power for their own purposes and those who are subject to its effects. Power, in this sense, may be used as an instrument of domination. (1996: 2)

A link is thus established between the differential capacities of actors, the summation of their powers, a tally of who gains and who loses, and domination in general. The apparent ease of this logic, however, sits uneasily alongside the earlier observation that the potential capacity for domination may also be understood as context dependent; that is, subject to prevailing conditions and circumstances. Part of the difficulty, as we shall see, rests with Hindess's insistence that any talk of parties holding more or less power brings to the fore an assessment of their quantifiable essence or capacity.

What is perhaps rather misleading is that Hindess should consider the term 'capacity' in relation to power as being almost univocal in its meaning. On this account, anyone who refers to power as a differential capacity, whether intentionally or not, is forced to endorse a quantifiable view of power. As such, the likes of Anthony Giddens and Michael Mann are reprimanded by Hindess for their interpretation of power as a set of capabilities which make it possible for individuals and organizations to achieve their desired goals. The fact that Giddens acknowledges that power can be conceived as a capacity to realize intended outcomes, and that such a capacity can 'make a difference' to eventual outcomes, is taken by Hindess as evidence of a simple, quantitative conception of power. Yet both Giddens and Mann in their writings conceive of power in a more rounded way than that, recognizing that power may have a collective as well as a distributional sense. Indeed, both writers recognize that power can be exercised *with* others, as well as *over* others.[2] Moreover, as will be argued in the following chapter, powers gained through association are not necessarily gained at the expense of others and are therefore less susceptible to simple 'more or less' equations of power.

Interestingly, where Hindess does make an exception to the sense of capacity as a quantifiable substance, it is the collective conceptions of power advanced by Hannah Arendt and Talcott Parsons which are drawn upon to illustrate power as a capacity bestowed by the consent of others.[3] Here it would seem that whenever the capacity of a particular authority to act is legitimized by consent, the quantities of power involved are less decisive to

---

2 Both Giddens (1977, 1984) and Mann (1986, 1993) recognize that power can be something that is mobilized to achieve collective ends rather than something inherent in social relationships which always works to the advantage of one party and to the disadvantage of the other. Indeed, both acknowledge the influence of the earlier work of Parsons on power as a 'generalized means' to further collective goals, although neither subscribes to Parsons's consensual approach to power. These points are spelt out further in chapter 3.

3 As indicated, Parsons's (1963) approach to power will be considered in the following chapter. Arendt's (1958, 1970) work will also be considered there, in particular her view of power as something produced by individuals 'acting in concert' to achieve common goals on the basis of shared understandings. It is indicative of Hindess's rather blinkered view of power as a capacity that Arendt's positive conception of power is acknowledged and her contribution briefly noted under the heading of power as a 'legitimate capacity' (Hindess 1996: 10–11).

the outcome. The assumption quietly at work here is that when power rests upon consent, it is less likely to be conceived as a property or an essence which is then capable of predetermining the outcome of distributive struggles. There is thus less call for a simple calculation of the quantities of power involved. The question never seriously posed by Hindess, however, is why, when there is talk of more or less power, this should automatically call forth a notion of capacity as a fixed essence capable of measurement.

The crucial point of all this is that the instrumental language of domination and 'power over' is far more flexible in its meaning than Hindess is prepared to allow for. It simply does not follow that to talk about groups or individuals who are more powerful than others is to ascribe to them automatically an 'essence of effectiveness'. Power as a capacity may be *attributed* to individuals on the basis of custom and practice, as Max Weber recognized; it can be *inscribed* as a causal force in oppressive and unequal relationships, as Karl Marx demonstrated; or it can be *implied* in a loose metaphorical sense, as one might say that advertising agencies 'hold' a capacity for seduction through their use of images and meaning. In each of these different senses of capacity there is nothing fixed or predetermined about the subsequent outcomes, and there is nothing to suggest that differential amounts of power are there simply to test our skills of addition.

Hindess, it would seem, considers that whenever power is referred to as a differential capacity there is some kind of trickery involved, whereby the exercise of power is merely the scripted expression of an inscribed capacity or essence. This is clearly unhelpful in the case of contemporary theorists of power such as Anthony Giddens and Michael Mann, who operate with a more differentiated conception of power. It is also radically misplaced to include reference to Max Weber as someone who subscribed to a conception of power as 'simple capacity'. More than most he understood the various combinations of ways in which power may be conferred and the various claims to legitimacy involved. Accounts which proclaim that power is a capacity possessed by individuals, groups and institutions alike can be interpreted in a loose, metaphorical sense or, at the other extreme, as an inscribed relational property. Only in the latter case should Hindess have any cause for concern, although even then quantitative scripts of power are not really what is at issue.

Potential capabilities and contextual outcomes

One of the most thorough accounts of power as an inscribed capacity is to be found in Jeffrey Isaac's *Power and Marxist Theory: A Realist View* (1987). Drawing upon the outlines of a realist philosophy of science developed by

Rom Harré and Ed Madden and Roy Bhaskar in the 1970s,[4] Isaac is probably the iconic figure that Hindess has in mind when he lambasts those who view power as a generalized capacity or, more pointedly, as an 'essence of effectiveness'. Implicit within this criticism is that power is conceived as some kind of unchanging, uniform substance which lies close beneath the surface of the daily life of individuals, social groups or institutions. If this were true of those who subscribed to a notion of power as inherent in the bodies of institutions, then this would indeed be a damning critique, but *contra* Hindess this is simply not the case.

Following Harré and Madden (1975), to attribute a power to any one thing is to say something about its constitution; that is, what it is capable of doing by virtue of its intrinsic nature. On this view, power is a causal property bestowed by the very structure of relations of which it is a part. Certain things are disposed to act in certain ways by virtue of their constitution, although whether or not such powers are realized is dependent upon prevailing conditions. Dispositions in this sense, to adopt a line of reasoning from Bhaskar (1975), may be possessed unexercised, exercised unrealized or realized unperceived, depending upon the context. Translated into the domain of social relations, it is the *enduring* nature of socially constituted relationships which Isaac locates as the basis of power and the source of an agent's capability:

> What I am suggesting is that a generalizing social science takes as its primary object of study those enduring social relationships that distribute power to social agents. Individuals certainly possess idiosyncratic powers. But what makes these socially significant is the way they are implicated in more enduring relationships. It makes perfect sense to claim that 'David Rockerfeller is a powerful man.' But a social theory of power must explain what kind of social relations exist and how power is distributed by these relations such that it is possible for David Rockerfeller to have the power that he has. To do this is not to deny that it is *he* who possesses this power, nor to deny those personal attributes determining the particular manner in which he exercises it. It is simply to insist that the power individuals possess has social conditions of

---

4 In particular, Isaac draws upon Harré and Madden's *Causal Powers* (1975) and Bhaskar's *Realist Theory of Science* (1975) and *The Possibility of Naturalism: A Philosophical Critique of the Contemporary Human Sciences* (1979). In the 1980s, Bhaskar went on to elaborate his particular version of transcendental or 'critical' realism in such books as *Scientific Realism and Human Emancipation* (1986) and *Reclaiming Reality: A Critical Introduction to Contemporary Philosophy* (1989). In the 1990s he developed a 'dialectical critical realism' (Bhaskar 1993, 1994) in which he distinguishes between power as a transformative capacity intrinsic to action and power as the capacity to get one's way despite resistance – that is, domination in general. See Andrew Sayer (1992) for a counter-argument to the accusations of essentialism within a realist philosophy of science.

existence, and that it is these conditions that should be the primary focus of theoretical analysis. (1987: 80–1)

For Isaac, then, it is the durable character of social relationships rather than their uniform or unchanging nature which endows an individual or group with power. There is no interest here in revealing a single, fundamental 'essence of effectiveness' in those who are said to possess power. Following Bhaskar (1975, 1986), Isaac is not seeking 'ultimate explanations' of the distribution of power, but rather fallible attempts to convey in succeeding accounts the enduring structure of relations which distribute capacities to some agencies but not others. As the nature and the make-up of groups and organizations change, as the power of institutions waxes and wanes, the task of analysis is to trace their altered social conditions of existence and to identify the new elements which constitute them. Invariably such a task is fraught with difficulty as an attempt is made to separate the potentially enduring characteristics from those of a less durable nature, but the aim is quite clear.[5]

Another way to think about what is meant by 'enduring' in this instance is to consider an example: that of the institutional power associated with the oil multinationals mentioned earlier, or with any of the software or electronics 'giants'. Multinationals, although not all big in terms of size, are involved in cross-border activities, and it is precisely because they are capitalized, multi-country operations that they have the power to switch investments from one location to another. Thus the ability to relocate operations or to exercise powers of acquisition and merger across national borders may be said to stem from the intrinsic nature of multinationals. However, to maintain such powers over time – and indeed space – a multinational engaged in the production of microchips, for instance, would not only have to remain profitable, but also have to be able to reproduce its own conditions of existence – in terms of technology, knowledge, capital and the like – often in new geographical locations. Power, in this sense, is said to be accumulated over extensive regions of space and time and deployed on the basis of 'reserves' which are continually reproduced by the firm. If such a firm fails to reproduce its capabilities or to develop others, then it loses its 'store' of power and thus ceases to be a multinational force.

5 Indeed, a distinctive feature of Bhaskar's work since *The Possibility of Naturalism* is the incorporation of a transformational model of social activity which stresses the mediated and reproduced nature of social structures. It is, of course, intensely difficult to establish the limits of mediation, although if anything there is a tendential determinism in Bhaskar's assessment of inscribed or structural powers which makes it difficult to talk about 'generated' powers in the way that Giddens (1984), for example, is able to in his theory of structuration.

Power, on this view, is clearly 'centred' in multinational institutions; it is something that they are considered to possess in relation to nation states or national workforces, although only for as long as they are able to reproduce themselves. It is not, then, an unchanging substance which sits just below the economic surface of firms, but rather a sustained capacity which is maintained in a wide, yet not unlimited, series of locations. Moreover, power is not conceived merely in terms of size, as if it were nothing more than a measurable component. Naturally there are 'big' multinationals and 'small' ones, which can be distinguished on the basis of the size of their assets, sales and employment numbers, but the difference in their capabilities is not a simple reflection of such quantitative characteristics. It is not merely the amount of capital at the disposal of multinationals which is at issue, but their technological and organizational capacities too, as well as the network of relations which constitute them. There is a qualitative side to this kind of power which is glossed over and ignored if it is presumed that power is merely a uniform substance capable of measurement.

It is obvious, I hope, from the above that the powers of multinationals, or for that matter the causal powers of any body or institution, can be held without being exercised or displayed. To speak of a 'reserve' or a 'store' of power in this context is thus to convey the impression that it is a capacity which may be drawn upon or used as and when the appropriate circumstances prevail. More realistically, as Isaac stresses, the successful exercise of such powers is a complex, contingent affair in a number of respects.

In the first place, particular institutions or groups with their own respective powers will stand in a multitude of relations to other agencies who may also possess a specific combination of powers. The specific powers of, say, an international finance house in New York, London or Frankfurt may rest upon a particular form of capital whose nature is to be liquid and mobile, but the ability of that house to achieve certain objectives will be enhanced, modified or deflected by the ensemble of relationships of which it is a part. Other financial and commercial institutions in those cities as well as in other international financial centres, together with their respective government bodies, will, through their actions, have a complicating effect upon the 'store' of powers held by any one institution and how they are exercised. In such a situation, neither the exercise of joint powers nor the annulment of particular capacities can be ruled out.

A similar set of contingent effects comes into play when the same international finance house extends the reach of its powers across national borders by distributing certain regulatory, disciplinary and risk-taking powers throughout its regional organizational structure. The delegation of powers from the centre, the distribution of power relations within the organization to achieve certain objectives in the mobile world of finance, is necessarily an open-ended affair simply because all outcomes are conditional. They are conditional upon the result of their interactions with other agencies and institutions in various global regions and also on the particular circumstances and conditions in which such powers may be realized. In short, there are always *other* capabilities and contexts to contend with.

Whatever credence we may wish to give Hindess's argument, it is evident that in Isaac's formulation of power as an inscribed capacity there is no sense in which the eventual outcomes of power plays are a scripted expression of some underlying essence. On the contrary, the intrinsic capabilities of roving multinationals or global finance houses are considered to be potential powers which may or may not be realized given the wide array of circumstances possible. It is in this sense that power, to stress an earlier point, may be referred to as dispositional in nature. There is always a likelihood or tendency for any individual, group or institution to act in certain ways because of the social relations which bind and reproduce them over space and time. Whilst such an analysis fails to address the scope and reach of such powers, concentrating instead on their distribution over space, it does none the less produce a 'thing-like' representation of power as a centralized capability which then radiates out through the surrounding social space. As such, there is little ambiguity as to its location.

It may be useful at this point to recall that there are other, more ambiguous ways in which power as a capacity may be represented, from the general idea that someone or some event has the capacity to change our lives to the somewhat firmer notion that power may be employed to advance certain interests at the expense of others. What is striking about this imagery, however, is that regardless of the tangibility of the representations involved, power as a capacity always seems to call forth a particular *modality* of power: namely, that of domination. This logic operates in Hindess's critique, but it is by no means peculiar to his assessment of power as a capacity.

Domination as a Mode of Power

Perhaps one of the main reasons for the deeply embedded association of power with domination in much social theory is that, Weber's influence notwithstanding, it is *the* plausible language of power. While Hindess readily talks about domination in terms of winners and losers, the word also invokes a sense of control, constraint, compliance or imposition in which the powerful are often counterposed to the weak. In the context of less stark oppositions, domination none the less conveys a sense in which the reach of multinationals, for example, or the apparatus of the state possesses the capacity to impose their will on others in a variety of situations. In all such cases power is conceived as a vertical relationship; power is something that is held *over* others.

Asymmetries of power

In making a case for the structural nature of 'power over' others, Isaac locates the asymmetry in the unequal distribution of power in society. As he puts it, 'relations of domination and subordination comprise a subset of power relations, where the capacities to act are not distributed symmetrically to all parties to the relationship' (1987: 84). Some people and some groups have more power than others, not by accident or by a series of fortunate events, but by virtue of the structure of relations of which they are a part. The capacity to secure advantage thus stands in relation to the potential loss realized by others who, in one way or another, are enmeshed in the same web of asymmetrical relationships.

Rather than a sporadic or random event in which domination takes place, therefore, the ability to secure the compliance of others can be seen in, for example, the vertical relationships of bureaucratic control exercised by managers and supervisors over those further down the chain of command; in the impersonal structures of political domination exercised by a unitary state over its peoples and territories; and in the relationships between global institutions such as the IMF and developing economies attempting to pull themselves out of poverty. In each of these scenarios, it is the relationship itself, rather than any particular actions taken by the dominant party, that potentially secures the compliance of the weaker party. For Isaac, this instrumental ability to direct and to constrain the practices of others is precisely what Weber had in mind when he spoke of domination as the 'authoritarian power of command'. Recruiting Weber to his argument, Isaac quotes approvingly one of Weber's more explicit definitions of domination:

> *domination* will thus mean the situation in which the manifested will (*command*) of the *ruler* or rulers is meant to influence the conduct of one or more others (*the ruled*) and actually does influence it in such a way that their conduct to a socially relevant degree occurs as if the ruled had made the content of the command the maxim of their conduct for its very own sake. Looked upon from the other end, this situation will be called *obedience*. (1987: 84; in original, Weber 1978: 946)

If the language of command and obedience, ruler and ruled, is taken to be the defining feature of domination, then clearly the parties tied to one another are, as observed, unequally related. Now it could be argued that because of the unequal distribution of power in society, it follows that any pay-offs to the parties involved in terms of privileges and benefits will be similarly skewed. This, you may recall, is the basis of Hindess's argument. While such an outcome is entirely possible, it is none the less far from inevitable. But there is a deeper concern lurking at the base of this assumption: namely, that domination is a zero-sum game.

A zero-sum game is one in which the scores of the winner and the loser sum to zero. There is only a fixed number of resources in play and thus one side's gain is another side's loss. In Isaac's account, despite the stress upon the contingent character of power plays, such a zero-sum game appear to frame the analysis. The successful exercise of power in vertical relationships, notwithstanding the manoeuvres performed by the different sides involved, would appear to benefit one side at the expense of another. Whether the loss or gain in absolute terms is one of power or one of benefits is unclear, but there is no mistaking the fact that, say, for workers to challenge management successfully or for the less developed economies to achieve some kind of progress, there has to be some give elsewhere in the (closed) system. In sum, the needs of the different sides are considered to be mutually exclusive.

Yet there is nothing about an asymmetrical view of power which requires that the pay-offs sum to zero. The outcome of power plays can be positive-sum, with no absolute winners or losers, or indeed negative sum, where losses of some kind are recorded by all players involved. In fact, all the unequal distribution of power signifies, as Weber well knew, is the *greater degree* to which one party may influence the conduct of others – nothing more, nothing less. Indeed, as Weber set down in his *Economy and Society* one of the most influential and closely reasoned accounts of domination as an asymmetrical mode of power, it is worth noting exactly what he did have to say on the matter, not least for the clarity of the ideas expressed.

Weber on domination

Significantly, for Weber domination represents only one among a number of different modes of power; it was for him a 'special case of power' which in its broadest sense involved the will of one party influencing 'that of the other even against the other's reluctance' (1978: 947). While there are any number of structures of domination shaping and influencing spheres of social action, from, as Weber noted, 'the social relations in a drawing room as well as in the market, from the rostrum of a lecture-hall as well as from the command post of a regiment, from an erotic or charitable relationship as well as from scholarly discussion or athletics' (1978: 943), the general characteristic of domination is that it involves some degree of *imposition and constraint*. It is important to be clear what Weber meant by domination here.

Situations of domination in this broad sense may involve relationships close at hand or across vast distances, but in either case the imposition of a form of conduct according to a set of particular interests is characteristic. The City of London in this general sense, as a powerful financial centre with a distinct and specialized position in the circuits of money capital, could be seen to influence the policy actions of the UK's political parties, departments of government, and industrialists alike in line with their own interests, despite the reluctance of the last to comply and without the slightest hint of obligation on their part. The language of influence in this context is thus far less restrictive than Isaac's selective interpretation of Weber's account of domination as command and obedience, ruler and ruled. There are no clear relations of interdependence involved between the City and other interested parties, and no formal lines of constraint, only 'objective circumstances' which realistically make it difficult for the other parties to do anything other than meet the interests of those in a dominant position. Above all, Weber stresses the fact that all parties involved are free to act in all kinds of ways, yet in following their rational self-interest they find themselves forced to fall into line. Although free in terms of their conduct, therefore, in practice those who are 'objectively' constrained have no choice other than to submit to the dominant will.

None of this is meant to imply that Isaac deliberately misinterpreted Weber's concept of domination. On the contrary, Isaac merely adopted Weber's 'narrow' or 'formal' sense of domination, which reflected Weber's own preoccupation with all matters of administration and organization in society. Reference to domination in terms of the power to command and the duty to obey represents a more structured, authoritarian form of organization which is tightly bound by rules and regulations, and in which there is little or no scope for the influence and interests of independent parties. Thus in contrast to Weber's broad sense of domination, in this more bureaucratic sense there is no question of one party making it difficult for the other not to comply with the first's wishes. On this restricted view of domination, there are no 'objective circumstances' which make submission the only realistic option. Instead, the asymmetry of relations is clearly inscribed in a vertical chain of command between superordinates and subordinates, where obedience is achieved through custom and practice or through the instrumental calculation of advantage.

Domination in this formal sense thus represents a more tightly orchestrated means of influencing the conduct of others. If constraint and imposition signify the core characteristics of domination, then *close discipline, continuous control* and *supervision* represent the organizational means by which domination may be achieved. The validity of such means, however, is by no means self-evident according to Weber, especially if domination is to be maintained on a continuous basis. For that to happen, such means have to be perceived as legitimate:

> Experience shows that in no instance does domination voluntarily limit itself to the appeal to material or affectual or ideal motives as a basis for continuance. In addition every such system attempts to establish and to cultivate the belief in its legitimacy. But according to the kind of legitimacy which is claimed, the type of obedience, the kind of administrative staff developed to guarantee it, and the mode of exercising authority, will all differ fundamentally. Equally fundamental is the variation in effect. (1978: 213)

So, if the exercise of domination in the formal sense is to be maintained, it has to be clothed in legitimacy. As is well known, this was the role Weber bequeathed to *authority*. Unlike domination, however, authority is conceded, not exercised in a straightforward sense. It is something that is claimed and, once recognized, serves as a means to secure a willingness to comply. It is not, as is sometimes said, simply a type of domination. To be sure, authority for Weber can be read as legitimate domination, but on closer inspection it reveals a particular means of securing assent within 'structures of domination' (1978: 941). Importantly for the argument of the book as a whole, domination should be seen as *one* among other modalities of power, one which can be used selectively in combination with other modes such as coercion to establish and maintain a structure of domination. In contrast to domination, the hallmark of authority, as Hannah Arendt pointed out, is 'unquestioning recognition', whose greatest enemy 'is contempt, and the surest way to undermine it is laughter' (1970: 45).

The distinction between domination and authority as separate but often related modalities of power also points up the difficulties of sustaining domination over time and indeed across great distances. The capacity to impose a set of personal or institutional interests on a continuous basis, potentially across different regions, is at best limited unless recourse to other modes such as authority is sought. One of Weber's preferred illustrations of precisely this point is the role that large central banks or major credit institutions can play in exercising a 'dominating' influence over the capital markets, in the first instance by virtue of their monopolistic position, and latterly by extending their 'powers' on a continuous basis through authoritative structures of control and regulatory supervision (1978: 943–4). In this way, such financial institutions may formalize the practices of domination by clearly setting out the obligations involved and the requirements to submit to them. Imposition in this context is considered to be particularly effective, because of not only the capacity of the institutions to exercise a general constraining influence on the conduct of the market but also the manner in which that influence may be consolidated through techniques of discipline and control and objectified in the form of authority.

Seen in this way, it is not altogether surprising that power, domination and authority are often treated as synonyms. In particular, the deeply embedded association of power and domination, however much we may pull them apart on reflection, has the ability, as Arendt warned, to reduce all kinds of power plays to the 'business of domination' (1970: 44). Authority, however, is not the same thing as domination, and as will be stressed throughout this book, neither is coercion, manipulation or seduction for that matter. Whilst there will always be some slippage of meaning between the different terms, the different modalities of power imply different forms of social relation which entail quite specific effects, as was argued in the introductory chapter. To conflate their different meanings, or worse, to reduce their specific effects to domination, is to misunderstand the diverse ways in which power achieves its effects.

Take the issue of cultural domination as it is often perceived in the case of the film industry or in advertising. The former, in the guise of Hollywood, is frequently portrayed as an agency of cultural domination. On a number of occasions in recent years, for example, French politicians have accused the US film industry of promoting a less than subtle form of Americanization: Hollywood images are said to promote uniform cultural tastes, guiding behaviour from the outside and spreading a particular cultural form like a sheen across the globe. Above all, they are perceived as a direct threat to French culture, an imposition on the French way of life. Clearly, there is a sense here in which the conduct of the French people is open to an external influence. The question, however, is whether or not such an influence is tantamount to domination. If it were domination, in Weber's general sense of the term, then the degree of constraint would have to verge on the monopolistic. Regardless of the percentage of box-office takings in France derived from Hollywood films, the US film industry would have to be in a position whereby it could 'prescribe' what is and what is not shown in French cinemas and on French television. In short, it would have to have a near monopoly on the French visual media through its control of the distribution channels which would make it difficult for audiences to view anything other than US products. If, however, French viewers on the contrary are not considered to have been stripped of all their autonomy, but to be free to opt for different visual fare, then the mode of power involved is closer to that of *seduction*.

Seduction, following Lipovetsky (1994), leaves open the possibility that a subject can opt out of the action. In contrast to domination, in either its general or narrow sense, seduction works at the level where choices are possible. It may encourage a desire for certain kinds of image and film, redirecting curiosity in the process, but it does so in a context where reflection and choice are both present. In the absence of cultural prescription, and no matter how gentle the methods of subversion involved, there is always the possibility of refusal or indifference. In making the case for advertising as a seductive rather than a dominating force, Lipovetsky stresses the limited effect of advertising as an *intended* effect:

> Advertising influences, but it does not threaten; it suggests, but it does not seek doctrinal domination; it functions without invoking Manicheanism or inducing guilt, in the belief that individuals can correct their own behaviour almost on their own, once they are aroused to responsible awareness by the media... Advertising does not take on the task of completely redefining the human race; it exploits embryonic tendencies that are already present by making them more attractive to people. Far from signifying an exponential race towards total domination, the spread of advertising reflects the reinforcement of a modality of power with minimal ideology and with strictly limited goals. (1994: 164–5)

Seduction as a mode of power, therefore, involves a renunciation of total domination, not its propagation; it is a modest form of power which is intended to act upon those who have the ability to opt out. To consider advertising as a form of cultural domination is then to misunderstand the ways in which the advertising industry seeks to influence social conduct. There is no orchestration of tastes along monopolistic or bureaucratic lines which disciplines the subject. Similarly, there is no sense in which *coercion*, the ability to influence conduct through the threat of negative sanctions, operates as an intended strategy of advertising. Obedience exacted through the threat of force, for example, is not conducive to advertising as a communicative medium. No more so, for that matter, is *manipulation*, where the concealment of intent serves to bring about the desired outcome. Whilst it is certainly the case that advertising may seek to influence by making veiled suggestions about its promotions or selectively restricting what is known about them, the process works on choices, on curiosity, not on an unwitting audience. The manipulation of needs is claimed rather than known to be an effect of advertising, as indeed is cultural domination. In considering the power of advertising or indeed the Hollywood film industry, therefore, it is instructive to note that domination is not the only mode of power held *over* others. There are any number of different ways in which one party may influence the conduct of others, with or without their resistance, but these ways are not all reducible to Weber's broad sense of domination or indeed to any looser sense of imposition.

Weber's preoccupation with a more 'technical' sense of domination, with the practices of discipline, control and supervision, reflected, as noted, his interests in bureaucratic forms of administration and organization. It was not intended as a general model of power, although he clearly argued the case that structures of domination are pervasive in the broadest of senses. While an emphasis upon the significance of different modes of power is central to the arguments of this book, it is, however, instrumental relationships of power, in particular that of domination, which inform most 'centred' conceptions. How that capacity to influence the actions of others is thought to be executed over space at distances far removed from the centre is one of the issues to which we now turn.

Centred Powers, Distributed Capabilities

As I see it, when a person or an institution, or a social class for that matter, is said to 'have' power, an implicit spatial vocabulary of power is invoked. Whether attributed to them, inscribed in them on the basis of structured relationships, or loosely implied, the possession of power itself identifies the location. The identification of so many different centres of power, however, does not exhaust that vocabulary. For Weber, and indeed Isaac, supervisory and regulating powers are distributed across organizations, for instance, on the basis of allotted roles and bureaucratic structures. Each person occupies a certain position in the vertical chain of command and each has a specific task to perform which can be executed only on the basis of orders received from above. The discretion that an official may exercise, in relation to others who have to carry out their demands or in relation to those on the receiving end of such demands, is strictly circumscribed. Strict adherence to the rules and procedures embedded within an institution is thought to avoid any elements of personal favour entering into the bureaucratic relationship.

In this case, the powers of the centre are said to be delegated through a variety of bureaucratic positions and realized in the form of administration. Domination in this context appears to be transmitted from the centre through a hierarchy of commands passed intact from the hands of one official to the next, so to speak. According to Weber:

> In the great majority of cases he is only a small cog in a ceaselessly moving mechanism which prescribes to him an essentially fixed route of march. The official is entrusted with specialized tasks, and normally the mechanism cannot be put into motion or arrested by him, but only from the very top. The individual bureaucrat is, above all, forged to the common interest of all the functionaries in the perpetuation of the apparatus and the persistence of its rationally organized domination. (1978: 988)

Not all forms of centralized power are transmitted in such a rigid, bureaucratic fashion, of course. But the sense in which power is extended or distributed *intact* is one that I would argue characterizes most 'centred' conceptions of power. When the capabilities of a middle-class group are extended to restrict movement across 'their' community, or a US film distributor attempts to disperse its capacity to control media outlets in Europe, or the financial and decision-making powers of a multinational concern trigger an effective capacity for control across a world-wide system of markets and production locations, their powers are assumed to be extended in such a way that the whole, unitary force comes into play. Expressed in different terms, the relationship between the centre and its extended powers is equivalent to that between a whole and a part. When a group or an institution extends its powers through the surrounding social space, it is *as if* the entire capacity is transmitted, made up as it were of a pre-formed power which is then delegated or distributed across the landscape. Once the capacity is known, whether awesome or otherwise, and extensive reach assumed, the rest of the equation quite literally amounts to a series of footnotes from the centre.

Footnotes from the centre

The spatial vocabulary of power here, then, is one of centres, distributions, extensions and delegated capabilities. It is as if a 'store' of centralized power is marshalled and transmitted intact through space and time and, allowing for an element of distortion and resistance, used to secure certain organizational or institutional goals. There is little in between in this linear story: success is attributed to the powers of the centre and its distribution, and failure is attributed to the extent of resistance. Whatever the outcome, the force of unitary, centralized power remains intact and its 'store' of capabilities present and awaiting distribution.

Because of the commonplace nature of this representation of power and space, it is perhaps not that surprising to find that it is frequently understated. It is for, example, one that underpins the 'state-centred' versions of power criticized by John Agnew (1994, 1998, 1999; see also Agnew and Corbridge 1995) in his attempt to map political power across state boundaries. His target was the conventional understandings of the geography of political power held by mainstream international relations theorists and those of a related political realist persuasion who considered the state to be a unitary and singular actor. At the core of his critique was the sense in which state territories had been reified as a fixed unit of secure sovereign space. In so far as convention distinguishes between an internal, domestic space in which governments exercise power in an orderly fashion over a defined territory and its peoples, on the one hand, and an external, international domain defined by the absence of order, on the other, the 'territorial state' in mainstream international relations is represented as a homogeneous political community maintained and controlled from an identifiable centre. The state as the central actor guarantees social order through the distribution of its powers to select elites and bureaucratic organizations and thus effectively 'contains' society within its territorial boundaries. As such, the spatial organization of rule-making authority is portrayed as an almost effortless process whereby power is impelled outwards from a capable centre.

There are parallels here with simplistic notions of money and capital which by virtue of some apparent inscribed qualities, a kind of capacity-in-waiting, are thought to exercise 'command over space' – as if it were simply a question of adding space to a pre-formed power.[6] Multinational firms can be thought about in this way, with their headquarter locations acting as centres of command and control in the global economy, co-ordinating and integrating the flow of capital movements and commodity markets across national boundaries. Although now something of a caricature, it is surprisingly easy to fall back upon this representation of economic power and total reach, where the multitude, dispersed and potentially far apart in territorial terms, are subject to the effortless reach of the 'big' transnational corporations. 'Command over space' is after all what domination is all about, but to secure and maintain it on an ongoing basis across far-flung locations was never going to be that straightforward. On its own, domination from afar, whether exercised through the operational HQs of multinational administrations or through the 'powers' of a unitary state over its sovereign territory, is always likely to be less than totalizing.

Much of the recent literature on economic and political governance, it seems to me, is precisely a response to this fact. Less boxed in by grand narratives and more responsive to what are perceived as the newly emerging sites or locations of power, there has been greater recognition among analysts such as James Rosenau (1997), David Newman (1999) and Bob Jessop (2000) that the exercise of power in a globally reordered world is less than straightforward. Increasingly, there is reference to 'multi-tier' or

---

6 Such a representation underpins much of David Harvey's earlier work on the capabilities of capital to dominate space. In *The Limits to Capital* (1982), for example, it is the mobilities inscribed within different kinds of capital which are seen to lie behind the uneven character of the built environment, and similarly, in *The Condition of Postmodernity* (1989), it is the capacity of capital to accelerate economic processes which is considered to have produced the latest 'round' of the annihilation of space by time, with the consequent domination of regional or place-based movements by the manoeuvrings of capital. Capital, as Harvey conceives of it, 'continues to dominate, and it does so in part through superior command over space and time' (1989: 238). See Allen (1997).

'multi-level' governance, where power is no longer seen to operate in either a top-down or a centre-out fashion, but rather upwards and downwards through the different scales of economic and political activity, both transnational and subnational. In such accounts, there is a greater recognition of the larger number of interests involved in any instrumental power formation, with multiple sites of authority dotting the political landscape, from numerous quangos and private agencies to local administrative units and other subnational actors. In place of the conventional assumption that the state is the only actor of any real significance, the playing field is now shared with NGOs, multinational enterprises and other supranational as well as interstate organizations. In this more complex geography, power is largely about the reorganization of scale, in so far as it is *redistributed* to take account of the proliferating sites of authority and reordered boundaries.

Globalization and the questions that it has raised concerning 'state-centred' approaches to power, in particular over the permeability of power in the face of the increased capacities of supranational institutions and its dispersal through quasi-autonomous regional and local agencies, have led to a more scalar vocabulary of power (see, for example, Brenner 1998; Swyngedouw 1997, 2000). Rather than a unitary centre of power, there are multiple centres, each with their own capacity to exercise authority and influence within the new territorial reordering. Rather than a simple top-down transmission of power, there is a redistribution or shift in capabilities between the different levels of governance. Above the nation state, for example, the agencies of the European Union, the regional outposts of multinational 'empires' and various NGOs are seen to exercise their influence over the actions of those within their realm, reaching down in many cases directly into the lives of those 'on the ground'. Similarly, within the nation state, the delegation and dispersion of power in the field, for example, of social welfare to a variety of trusts, partnerships and community organizations have increased the number of authoritative locations with responsibility for ordering people's lives.

And yet, for all this talk of power shifting between the different levels or scales of activity, the story remains a remarkably linear one. Multi-level governance is largely about a descending order of spatial scales – from the supranational and the national down through the regional to the local, where each scale appears as unified and as ordered as the 'territorial state' once did before it. Where once it was the state and its territories which were reified as a fixed, 'contained' unit of space, now it seems as if the different levels of governance have inherited this geographical mantle. Despite recognition of the fact that the workings of power involve more than a simple vertical or horizontal reallocation over space, there appears to be a kind of scalar mindset in operation whereby the region is thought to dovetail into the nation, the nation into the European Union and so forth. In this revised linear story there is much more to consider than the powers of a unitary centre and their remote impact, yet the vocabulary of power employed is still one of capabilities 'held' and the dispersion or delegation of powers between the various levels and sites of authority.

In this territorially reordered set of spaces, it is still hard to avoid the impression that power is conceived as something which is more or less transmitted *intact* across the scalar landscape. There may be more centres of power and authority to consider, more complex dispersals and delegations to contemplate, but what happens to those on the receiving end may still be read as a series of footnotes to the (by now multiple) centre's capabilities.

Out-of-Scale Images

In *Culture and Imperialism* (1994), Edward Said spoke about 'out-of-scale' images, images which distort our understanding of the significance of something and what it is capable of bringing about. Said had the images of 'terrorism' and 'fundamentalism' in mind, 'which derived entirely from the concerns and intellectual factories in metropolitan centres like Washington and London' (1994: 375). I am reminded of the image of mammoth economic corporations roaming the globe, or of the colossal influence of such worldly institutions as the World Trade Organization or the IMF, or, closer to home, of the penetrating influence of state agencies deep into the fabric of everyday life, each of which in its own particular way creates an overblown sketch of what these organizations are capable of bringing about. Such images, related to size but also to their imagined capacities, distort our sense of power's reach. It is as if the very images themselves signify a reel of effects which can be far reaching or established close at hand over the lives of those dispersed in diverse and often distant locations. These out-of-scale images are fed by a spatial vocabulary of power which talks of power's extension (or command), over space, its redistribution across a deeply contextualized landscape. Curiously, it seems that the very fact that power may be thought of as something which can be extended or distributed over space actually makes it that much harder to conceive that it may, in practice, be *actively constituted in space and time*.

But, perhaps more than that, it is the language of power as capacity which sets in motion the inflated sense of power's effortless reach. For me, this is misleading, as the so-called 'capacity' of power is often a euphemism for the resources and abilities which may or may not be mobilized to produce an effect – be it domination, authority, seduction, manipulation, coercion, inducement or whatever. Resources move and can be lost or may simply evaporate, but power has none of those qualities. The relationships which give rise to the constraints of domination or the allure of seduction or the assent which authority requires are mediated through relations of proximity and presence. Power is not some 'thing' that moves, but an effect that is mediated, and such effects may mutate through relations of successive or simultaneous reach. But that is to run ahead of ourselves.

The idea that power may be something other than a capacity, or that power may not be so much distributed over space as constituted by the many networked relationships which compose it, has yet to be fully absorbed by those who remain convinced that power may be possessed or held 'in reserve'. For that understanding, we need to draw on a different conception of power, one that starts from the position that power is an effect, not an attribute of 'things'.

3

Power through Mobilization: From Mann's Networked Productions to Castells's Networked Fictions

In contrast to the more familiar idea of power as something that is held or possessed, abused even, the conception of power developed in this chapter is far less intuitive. It is, after all, far less obvious to think of power as a medium which is brought into being through the mobilization of collective or individual resources than to accept that it is something which is held in reserve and hung on to at all costs. On this view, power is generated to achieve certain outcomes, and thus it makes little sense to talk of power as 'contained' within things or 'stored' ready for use. If power is an effect which is produced through the actions of groups or individuals, then it is not something which may be held in reserve: it can only be *mobilized* on what is often a loose and tenuous basis.

Power, on this account, far from appearing solid in form, is thus represented as a rather fluid, amorphous medium which may evaporate just as easily as it may fill out. The ability of people to define themselves in terms of dress or bearing, for example, and impose this definition upon the minds of others, rests upon their ability to mobilize resources – symbolic in this instance – and to use them as an expression of their particular interests. Where no such collectivity existed before, their very constitution comes into being through the power of naming, of designating, of making people see and believe in a world where they exist in contrast or in opposition to others groupings (following Bourdieu 1989, 1991). In this way, groups, whether conceived in terms of class or constructed around other stylistic possibilities relating to place, ethnicity, sexuality or whatever, may actively empower themselves through the collective mobilization of resources. Equally, however, they may lose their power the moment that they fail to act together or diminish its effect once collective, short-term goals have been won. Crucially, there is no locatable 'reserve' or 'store' of power available to be tapped when needed.

Having said that, power, on this view, need neither be associated with particular interests, nor tied directly to the practices of opposition and domination. The mobilization of resources may take place, as Talcott Parsons and Anthony Giddens have both argued, in the context of simply enabling things to get done. Power here is seen as facilitative, or rather transformative, in so far as it 'makes a difference' to eventual outcomes. In the context of a wide range of institutions, from the governance of schools and hospitals to the running of multinational enterprises, power may be thought of as more a means to achieve outcomes, and rather less an instrumental ability to get one's way despite resistance. Power, as such, is not considered as merely something which is capable of being held over others; it may also be enabling. In the hands of Hannah Arendt, this understanding of power, as we shall see, was considered as empowering in and of its own right, as a positive gesture in which all those involved benefit in some way, yet only for as long as the effective mobilization lasts.

In the light of all this, it is perhaps not altogether surprising to note that such a fluid medium is relatively difficult to locate, especially in comparison with power which is said to be 'centred' in social bodies. Significantly, it is the instantiation of power in social action which gives it the appearance of an amorphous entity. Its lack of shape does not mean that it lacks spatial definition, however, since it is understood that power is produced through networks of social action and indeed may expand or decrease in such networks. It is this process, the mobilization of resources through such networks, which provides the basis for a particular spatial vocabulary of power.

In this chapter, this spatial vocabulary is explored through Michael Mann's prodigious work, as well as that of Manuel Castells, where a spatial lexicon is more or less explicit, although in other more resource-based versions of mobilized power it is often only barely traceable. Even where acknowledged, however, it is the notion that power is something that may be generated at different sites and locations across the networks and projected unproblematically between them that, I argue, loses sight of the difference that spatiality makes to the exercise of power. It is simply a fiction to believe that power flows merely because it is a 'fluid' medium.

However, before we consider the relationship between spatiality and networked power, it is important, to be clear about what is meant by power as a *medium*.

Power as a Medium

In referring to power as the means to achieve outcomes, it is relatively easy to slip into an understanding that power is itself a specific type of conduct or action. It may not be thought of as residing in anyone or anything, but it may be characterized as a specific act when, for example, we say that a person of a particular manner or bearing 'exudes' power, or that a particular social group has been known to 'exert' its power in the marketplace. In such circumstances, it may appear possible to identify when someone or somebody is acting in a powerful manner and when they are not. The sense in which power is a medium, however, is not about isolating a range of actions or forms of conduct which are readily identifiable as exercises in power. Rather, it is about identifying how power is produced *in and through* social interaction.

To return once more to the example of the middle classes and their assumed ability to divide up the geography of cities through the processes of exclusion and spatial closure: it is possible to conceive of such powers as something generated as an effect of the middle classes' collective association. As such, their abilities may be seen to derive from the particular ways in which people pool their resources, both material and symbolic. Power, on this view, is produced through people acting in association: through the individual and collective resources which they bring to bear on the situation. In this context, it is possible to see how resources such as money, contacts and information, as well as other social skills and symbolic competencies, when used or pooled collectively generate an effect which may be called power. This quality is not some kind of latent capacity possessed by the middle classes to exclude from their immediate surroundings those that they fear the most, but simply a means produced through the mobilization of resources to achieve such an outcome.

Interestingly, one of the most significant accounts of conceptualizing power in precisely this manner was first put forward by Talcott Parsons in a series of articles in the early 1960s, in direct response to C. Wright Mills's fundamentally Marxist, 'centred' view of power, expressed in his book *The Power Elite*, published in the late 1950s.[1] Although not known for his subtle theorizing, Parsons's account of power offers a number of specific insights on power and its constitution.

---

1 The debate between Parsons and Mills – if we can call it that, as Mills never responded directly to Parsons's critical review of *The Power Elite* (1956) and Parsons himself developed his ideas only after Mills's death in 1962 – is significant for the elaboration of a vocabulary of power which contrasts the 'power to' act with the exercise of 'power over' others (see Parsons 1957, 1963). The distinction between power as

More or less power

One of Parsons's central concerns was to demonstrate how power can be seen to act as a symbolic medium in society, in much the same way as money acts as a medium for exchange in the marketplace. The analogy with money is intentional, for Parsons wishes to make the claim that power, like money, owes its significance not to any intrinsic properties it may possess, but to its institutionalization as a social symbol. Money is able to circulate freely as a medium of exchange because of the confidence that people attach to its 'value'. Whether involved in the purchase or the sale of commodities, both parties to the transaction perform the exchange in the knowledge and expectation that the money they hold will be recognized by others in similar contexts and may later be 'cashed in'. Such compliance is deemed to be binding on all parties involved. Thus, in so far as money enables exchange to take place without the equivalent back-up in gold, so, Parsons argues, does power oblige others to get things done without recourse to threats of violence or coercion. Symbolic power, as Bourdieu among others recognized, is almost magical in quality in that its ability to mobilize rests solely upon others recognizing and legitimizing its form.

Or in Parsons's more disenchanted tone:

> just as a monetary system resting entirely on gold as the actual medium of exchange is a very primitive one which simply cannot mediate a complex system of market exchange, so a power system in which the only negative sanction is the threat of force is a very primitive one which cannot function to mediate a complex system of organizational co-ordination – it is far too 'blunt' an instrument. Money cannot be only an intrinsically valuable entity if it is to serve as a generalized medium of inducement, but it must, as we have said, be institutionalized as a symbol; it must be legitimized, and must inspire 'confidence' within the system – and must also within limits be deliberately managed. Similarly power cannot be only an intrinsically effective deterrent; if it is to be the generalized medium of mobilizing resources for effective collective action, and for the fulfillment of commitments made by collectivities to what we have here called their constituents; it too must be both symbolically generalized and legitimized. (1963: 243)

constraint and power as facilitative, although not without its philosophical precedents, was drawn by Parsons in response to Mills's assertion that the US in the 1950s was dominated by a centralized economic, political and military set of interlocking elites. The distinction proved fruitful at the time within sociological circles and, as shown in this chapter, was picked up by Giddens and adapted to his particular concerns and interests. More recently, it is the positive, enabling side to Foucault's conception of power, outlined in the following chapter, which has rekindled interest in the 'power to' act as part of what it means to 'govern the self'.

In considering the fact that power has to circulate through society, flowing, as Parsons puts it, back and forth across the boundaries of the polity, a high degree of shared understanding and confidence in those who hold authority is thought necessary. Leaving to one side the issue of how far such a degree of legitimacy is likely to be conferred, the idea that power flows as a circulatory medium, analogous to money, is used by Parsons to demonstrate the limitations of a zero-sum conception of power. In an attempt to stress the facilitative side to power – that it may be generated to achieve collective goals rather than sectional interests – the circular flow of power is shown to be capable of enhancement in much the same way as interest payments enhance a sum of money invested.

Anthony Giddens neatly summarizes Parsons's thinking on this matter in an early review:

> In Parsons's view, the net 'amount' of power in a system can be expanded 'if those who are ruled are prepared to place a considerable amount of trust in their rulers'. This process is conceived as a parallel to credit creation in the economy. Individuals 'invest' their 'confidence' in those who rule them – through, say, voting in an election to put a certain government in power; in so far as those who have thus been put into power initiate new policies which effectively further 'collective goals' there is more than a zero-sum circular flow of power. Everybody gains from the process. Those who have 'invested' in the leaders have received back, in the form of the effective realization of collective goals, an increased return on their investment. It is only if those in power take no more than 'routine' administrative decisions that there is no net gain to the system. (1977: 335)

If the outcome of such a scenario can be positive-sum, however, it can also be negative-sum, with those who have 'invested' resources losing out as the system spirals down into the equivalent of an economic crash. Arguably the analogy with the circuitry of money can be pushed only so far (see Barnes 1988), especially as many of the rather elaborate parallels drawn by Parsons between money and power appear ostensibly to rely on the mechanisms which enable the former to circulate rather than the latter. The abstract movements of money do not translate that easily into the idea that power is also 'deposited', 'brokered' or 'credited', or to the idea that an exchange of powers will occur in a smooth and stable manner. None the less, as Giddens (1977) was quick to recognize, Parsons's analysis of power as a medium circulating or flowing through a variety of social arenas afforded two significant insights.

The first concerns the fact that power may be conceived as a *fluid medium*; it can expand in line with the resources available to collective ventures, or it can diminish once collective short-term goals have been achieved, or, if we extend the insight, it may disappear once alliances fall apart. Power, on this view, is not restricted to a fixed amount of resources in play; it is capable of being generated over and above whatever resources are mobilized. Whilst Parsons's own illustrations of this process are rather limited, it is not difficult to conceive of the middle classes coming together to enhance their powers over and above their individual abilities, or indeed of such powers evaporating, should their resource base be undermined in a particular area. Much the same can be said of the formation of social movements, in relation to the environment for example, or indeed of firms teaming up with one another to exploit a particular leading-edge technology rather than risk going it alone. The possible scenarios for collaboration are numerous and far in excess of Parsons's systemic examples of liberal democracy 'in action'. The point is that the amounts of power in circulation are, quite simply, not fixed but variable and context-bound.

The second insight concerns Parsons's stress upon the fact that power in circulation may be used to the advantage of all concerned; it is a *general facility* for achieving a range of goals which are held in common, rather than directed towards the realization of particular interests. Whilst Giddens is rightly critical of Parsons's indifference to conflicts of interest in society and the power plays which lead to constraint and domination, he is keen to preserve the sense in which power 'becomes an element of action, and refers to the range of interventions of which an agent is capable. Power in this broad sense is equivalent to the transformative capacity of human action: the capability of human beings to intervene in a series of events so as to alter their course' (1977: 348).

On this view, power is intrinsic to all social interaction; it is the medium by which events are transformed, regardless of whether or not such an outcome serves one particular set of interests over another. Power changes things and power in circulation holds out the possibility of transformation across a wide variety of social arenas. As a fluid medium, the generation of power through the mobilization of resources in different contexts may enable things to get done to the satisfaction of all parties involved, or to the benefit of one side at the expense of another, or indeed to the advantage of the collective-minded over their less collective rivals. In Giddens's hands, divest of its consensual overtones, power is no longer simply 'in circulation'; it is a medium more or less generated over variable spans of space and time.

This is not, it should be stressed, simply a concern with the extension of some centralized power over space, be it the inherent capacity of the nation state or that of a particularly 'dominant' social group. Rather, what Giddens has in mind is a conception of power *produced* through social interaction, which itself stems from and in turn promotes the 'stretching' of social relations over space. This view is similar in many respects to Michael

Mann's conception of power as a series of overlapping, intersecting networks, in so far as both he and Giddens refer to power as the mobilization and deployment of resources over space. Although their spatial vocabularies differ somewhat, their understanding of how power transcends the boundaries of space and time has much in common.[2]

Resource Mobilization and the Production of Power

Both Giddens and Mann begin their analyses of power from the same vantage point, namely the perception that power rests upon the utilization of resources. Crucial to this assumption is the recognition that power itself is *not* a resource, but rather something generated or actualized through the control and reproduction of different kinds of resources. The power simply to do things or to influence the actions of others may be enhanced through the employment and application of resources, be they economic, informational or military, over tracts of space and time. For Giddens at least, the mobilization of resources over space and the ability to control them represents a *distanciated* form of power. The more that social relations are affected by what happens elsewhere, the greater the extent to which social activity in one place is governed by relationships elsewhere, the more powerful is the means to transform events. On this view, power is generated and its expansion made possible through the ability to control events over space and time. If societies are more 'stretched' than hitherto, because of advances in modes of communication and information, then power is both constitutive of that process and, in turn, enhanced by it. This is a rather truncated version of the argument, so let us return to what Giddens means by resources and how they are mobilized over space.

In *The Constitution of Society* (1984) and in his earlier writings, Giddens distinguishes between allocative resources and authoritative resources. By 'allocative resources' he means what, at first sight, appear conventionally as material resources, such as property, land, goods, technology, access to finance and the like. Authoritative resources are of a different order and refer, as the term implies, to the authority or control over the ways in

---

2 Although Pierre Bourdieu (1989, 1991) also considers power as a medium through which social groups mobilize their collectivity in and across social space, his spatial vocabulary of power, as hinted at in the chapter introduction, draws upon the use to which distance and propinquity are put in the construction of group formations. Interestingly, he makes a sharp distinction between geographic and social space, where the former is defined by fixed distances and precise segregations rather than the symbolic dexterity of the latter. See also Painter (2000) on the distinction. I return to accounts of social space in chapter 7, where the rhythmic and routine construction of social space in the interests of so-called dominant groups is considered at length.

which social life is organized and distributed over space. Neither type of resource is fixed in the general sense of the term and the two types are to be found in different combinations and relations, although the expansion of material resources is said to require a transformation in the organization of authoritative resources. Significantly, however, both types of resource refer to sets of capabilities rather than to material objects or people as such. Resources, as Giddens insists, are 'the media through which power is exercised' (1979: 91); they are not possessed by individuals or groups but drawn upon in the course of social interaction. 'Allocation', as Giddens argues, 'refers to man's capabilities of controlling not just "objects" but the *object-world*... Authorization refers to man's capabilities of controlling the humanely created world of *society itself*' (1981: 51). As such, resources are bound up with the systems of shared meaning which make it possible for the control and co-ordination of material and social life to take place.

At its simplest, if power is the means to achieve outcomes, then such an aim will inevitably depend upon an ability to influence the actions of others. Like Parsons's argument, however, this one is conducted at the level of social systems, where social relations are constituted as a systemic practice. So for resources to work as a structural element of social life, they must be taken up and reproduced through social practices on a routine basis. In different epochs, in different types of society, this process of reproduction will take different forms with different amounts and kinds of power generated. People comply, or rather their conduct is influenced, not in response to the 'wishes' of others, but in response to those who have the most 'leverage' over material and social resources.

How this 'leverage' occurs in relation to the embedded structuring of practices and resources in modern and not so modern societies is not altogether clear, although for Giddens the 'storage' of resources does appear to hold the key. The term 'storage' is perhaps a little unfortunate here, given its tactile, rather solid connotation. However, Giddens is not referring to a 'reserve' of power which may be tapped when required, but to a concentration of resources which serve to 'bind' or stabilize sets of social relations. As the amount of resources – property, information, technology, knowledge, modes of organization and the like – at any one location expand, so too does the 'leverage' of power and its extension in space and time. According to Giddens, just as city-states once acted as crucibles for the generation of power, as 'containers' of power in their own right, eventually losing out to the territorially bounded nation state under the impact of wider market relations, so today the nation state is under threat from even more 'disembedded' forms of social interaction.

The main point here is that resources and their 'containment' across time-space are considered to be the principal means by which power is produced and which, in turn, extend the 'leverage' across ever greater distances. As control over longer spans of time and space is achieved, so the means of power available is directly enhanced. In Giddens's account of the process:

> The storage of authoritative and allocative resources may be understood as involving the retention and control of information or knowledge whereby social relations are perpetuated across time-space. Storage presumes *media* of information representation, modes of information *retrieval* or recall and, as with all power resources, modes of its dissemination. Notches on wood, written lists, books, files, films, tapes – all these are media of information storage of widely varying capacity and detail. All depend for their retrieval upon the recall capacities of the human memory but also upon skills of interpretation that may be possessed by only a minority within any given population. The dissemination of stored information is, of course, influenced by the technology available for its production...It is the containers which store allocative and authoritative resources that generate the major types of structural principle in the constitution of societies. Information storage, I wish to claim, is a fundamental phenomenon permitting time-space distanciation and a thread that ties together the various sorts of allocative and authoritative resources in reproduced structures of domination. (1984: 261–2)

On this view, then, if today the power 'contained' within the nation state is beginning to evaporate, it is primarily the reorganization of informational resources across large time-space distances which pose the greatest threat. As expressed in *The Consequences of Modernity* (Giddens 1990), globalization, as a kind of shorthand for a range of disembedded relations and institutions, from the money markets and transnational firms to expert systems and communications technologies, has the ability to 'lift' social relations out of nation state involvement. In a world of rapid, intensive information flows, the mobilization and retrieval of resources over space are likely to be in the control of those disembedded institutions which are capable of linking local practices to global social relations. In this way, more distanciated forms of power involving interaction with those physically absent may be seen to represent a modern, facilitative means of securing outcomes. Action 'at a distance', Giddens argues, may be thought about as both enabling of social action and also constraining of it, as for example in the case of surveillance, where the increasingly sophisticated 'storage' and monitoring of information, images and sounds may lead to the detailed control of people's daily lives (see Giddens 1985 on the assumed role of the nation state in this process).

Thus, on this understanding, power is clearly not some pre-formed entity with latent abilities. Rather, distanciated forms of power may be considered to manifest their abilities through a series of often routinized and repetitive practices 'stretched' over space. Administrative power, in transnational institutions for example, can be seen to work through an ability to regulate the timing and spacing of social activities. Those 'locales' in which resources tend to concentrate, in cities, nation states and beyond, are said to 'contain' sufficient powers to control events further afield. What may also be understood from this, however, is that power generated in one part of a distanciated network is transmitted intact across it. As an understanding of power, therefore, it is possible to read such distanciated networks merely as *conduits* for the transmission of all kinds of organizational and institutional ability. And, somewhat surprisingly, these networks seem to undergo little in the way of displacement or transformation.

For a more penetrating account of how the process of resource mobilization is actually organized and controlled over space, however, we need to turn to the work of Michael Mann.

Networked (re)sources

Mann's overarching concern is with the organization of power, the infrastructural means by which people and territories have been effectively controlled by institutions and social groups of whatever hue since time immemorial. In his two-volume *The Sources of Social Power* (1986, 1993), a selective mapping of the logistics of power is attempted which focuses on the flow of information, ideas, people and materials across historical borders and territories. The mapping is selective in so far as it only considers those locations and states which in some sense represent what Mann refers to as the 'leading edge of power'; namely those that exhibited new and novel extensive and concentrated organizational means. From Mesopotamia in the sixth century BC, through Phoenicia, classical Greece and imperial Rome, up to the emergence in western Europe of the nation state and classes as the major actors of modern times, he seeks to trace the developed 'infrastructures' of power related to each historical moment. (A third and fourth volume are planned which bring the analysis into the twentieth century and provide a broader set of theoretical conclusions.)

Following Parsons, Mann acknowledges power as a means to secure outcomes, generated by the mobilization and deployment of resources, or rather 'power sources', as he prefers to call them. The distinctive twist in his approach, however, comes from the recognition that the mobilization and control of resources actually takes place through various *networks* of extensive and intensive social interaction. In brief, the expansion of power and its consolidation are said to take their shape from a series of networks organized over space which cross-cut and overlap one another, the most important of which stabilize around four types of (re)sources – broadly defined as economic, ideological, political and military. The result is a curious blend of Parsons on power, Weber on organization and control, and Mann on overlapping networks.

Power networks in this context take on a fairly conventional meaning. At their simplest, they are formed through patterns of association and interaction which bind people and bodies together in the pursuit of certain ends. A wide array of institutions and practices, from the broad, sweeping alliances of geopolitical institutions and their equally international economic counterparts, transnational firms, at one end of the spectrum, to the more regional associations of culture, religion or political practice, at the other, connect people and places together over shorter or longer distances. Differences in the make-up and dynamism between networks of interaction ensure that they reach out across space in different ways and to varying extents, in some cases transcending established social boundaries and in others heightening or consolidating them. Among the most powerful networks, those centred on the most effective (re)sources, a more stable shape and distinctive form of organization are assumed to emerge. In laying one network over another, however, as in some kind of lace pattern, the dynamism of each is said to fuse and modify the other's pattern of interaction so as to bring forth all manner of unintended and unanticipated consequences. The result for Mann is a view of history as a complex 'mess' and a geography of social relations which continually eludes simple notions of societies as nationally bounded entities. In his second volume, Mann favours the metaphor of 'entwinement' to capture the way in which the different sources of power 'change one another's inner shapes as well as their outward trajectories' (1993: 2).

Leaving the 'mess' that is deemed to be history to one side, a key consideration in all this is Mann's discrimination between the organizational abilities of different types of power to achieve their goals. Taking a leaf out of Weber's writings on power and organization, Mann distinguishes between authoritative power and diffuse power and relates them broadly to the manner in which different types of resources are organized and mobilized:

> *Authoritative power* comprises willed commands by an actor (usually a collectivity) and conscious obedience by subordinates. It is found most typically in military and political power organizations. *Diffused power* is not directly commanded; it spreads in a relatively spontaneous, unconscious, and decentred way. People are constrained to act in definite ways but not by command of any particular person or organization. Diffused power is found most typically in ideological and economic power organizations. A good example is market exchange in capitalism. This involves considerable constraint that is yet impersonal and often seemingly 'natural'. (1993: 6)

The similarity to Weber's distinction between domination in its broadest sense as imposition and constraint, on the one hand, and its narrower, more formal meaning as discipline and control, on the other, is not coincidental. As with Weber, the powers of economic organizations involved in market exchange are regarded as diffuse in form and extensive in their reach. In contrast, the powers of the armed forces or those of the state are seen as authoritative, based upon the rules of command and obedience, and although concentrated in form they are considered quite capable of extensive penetration. In focusing upon the *spatial reach and scope* of the different networks, however, Mann is adding a further dimension to an understanding of their organizational means; namely, their geography.

In military networks, for instance, power is regarded by Mann as essentially concentrated, coercive and highly mobilized relative to place. As the social organization of physical force, military power is seen as intensive, tightly controlled and able to command a high level of commitment from all those mobilized. When stretched over large expanses of territory or enacted at great distances, however, coercion is limited in what it is able to achieve and control of daily life is largely restricted to the production of compliance. The logistics of military power, that of 'concentrated coercion', are argued by Mann to be quite different from those which characterize state political power, however. Here power, although organized on a far more centralized basis than is common to military alliances, is assumed to be capable of extensive reach into the daily lives of its citizens. Through the generation and concentration of 'infrastructural' resources – customary taxation, property administration, market regulation and the like – the state is able to 'penetrate society and to implement logistically political decisions' (1986: 170).[3] Despite such differences in their organizational reach, however, neither military nor state power is conceived by Mann as diffuse in form. The spatial reach of economic organizations, national and international firms, as noted, is said to adopt this form, whilst remaining

---

3 Mann (1984) develops the distinction between despotic and infrastructural power, where the former refers to the powers of state elite groupings to impose their will regardless of the interests of civil society, and the latter to the penetrative reach of the state into the furthest recesses of their territories. The significance of the latter, for Mann, is that it is through infrastructural power that the state is able to hold on to its territories as 'resources radiate authoritatively outwards from a centre but stop at defined territorial boundaries' (1984: 198). How this 'centred' view of political power (which equals the state's organizational power for Mann) with its supposed unified reach sits alongside his networked forms of mobilization is not entirely clear, although this kind of theoretical eclecticism characterizes his work. It is interesting to note that within the field of political geography his work has been picked up less for its geographical insights than for its classification of (re)sources of power (see Painter 1995; Robinson 1996; Muir 1997; Ó Tuathail 1996).

capable of both intensive control (of the workforce) and extensive market constraint should near monopoly conditions apply. Perhaps the most intriguing example of diffused power offered by Mann, however, concerns the mobilization of cultural resources – the construction and regulation of shared meanings, the interpretation of ideas and images, and the spread of social practices and rituals – through the networks of cultural and ideological organizations.

Less organized in a systematic way than either state political or military power, the effective mobilization and control of cultural resources is said by Mann to rest upon persuasion rather than manipulation, and certainly not on coercion. In the previous chapter, I spoke of manipulation as involving the concealment of intent or the selective restriction of information. By way of contrast, persuasion as a modality of power is similar to seduction, in so far as it works at the level where choices are possible and is at its most suggestive where existing concerns and interests are projected as desirable. Persuasion as a form of social interaction does, however, to my mind lack the affective resonances that Lipovetsky (1994) ascribes to seduction, namely the ability to bring into sharper definition the aspirations and wants of all kinds of different people. In fact, the kinds of nationalist movement, religious movement and class solidarity which Mann has in mind when he speaks of ideological persuasion would appear to require the intensity of seduction if any reasonable hold over people's 'hearts and minds' is to be attained.

None the less, what Mann clearly understands is that cultural beliefs and shared understandings are capable of extensive reach. The diffusion of ideas and cultural practices, though word of mouth, travel, migration, images and all manner of publications, is capable of transcending economic and political borders and mobilizing far beyond them. The easy flow of ideas, however, is not be confused with their cultural embeddedness, which is likely to be of lower intensity the greater the distances involved, especially where the cultural contexts differ markedly. It is, after all, one thing to point to the spatial diffusion of cultural beliefs and practices and quite another to know how they are received, interpreted and translated. The effects of cultural or ideological power 'at a distance' are, I would have thought, far from transparent.

Indeed, the absence of any monolithic power associated with a particular resource is more or less understood by Mann when he notes that the most effective institutions in terms of power are those which encompass all four forms of organizational reach. An understanding that powerful institutions may enhance their means through extensive and intensive networks, combining both authoritative and diffused techniques of organization to achieve far-flung goals, is precisely the distinctive contribution that Mann makes to an understanding of power as a medium (notwithstanding his tendency to treat institutions and resources as of one, despite an 'ideal types' disclaimer). In fact, it could be argued that his singular achievement is that he provides an insight into how it is that distanciated forms of power actually mobilize resources 'at a distance'. More recently, he has brought this insight to bear on the differences in the make-up and dynamics of globalization as an historical phenomena. Speaking about the blend of the different sources of power which comprise and, more pointedly, shape the contemporary global landscape, he charts the mixed or rather 'entwined' pattern that has become his intellectual trademark:

> Northern capitalism unevenly but simultaneously integrates, dominates and ostracizes across the world. The power of the military hegemon, the United States, is somewhat undercut by its own increasing pacification; by its reliance on supposedly safe but decidedly blunt long-range weaponry; and by new weapons of the weak. Political power remains primarily wielded by nation-states, though some of these are more stable than others. Ideological power expresses all these diverse relations. Such complexity is not new to human societies. Globalization merely changes its scale. (2001: 72)

The sweep of the analysis is, as ever, impressive, but in formulating such generalizations there is, for me, an emptiness at the heart of what is actually involved in the implied extensive reach of such complex and diverse power relations. As I see it, *prescribed* differences between power (re)sources rather than the spatiality of power are what, for Mann, predispose networks to operate in the ways that they do. Whilst his grasp of the logistics of power (re)sources and their organization is admirable, there is a real sense in which, for him, networked social relations merely act as the *carriers* of power through their ability to transmit organizational resources. We will return to this point shortly, but before that there is another distinction of power to consider.

Powers of Association

At various moments in this chapter we have shifted back and forth between two rather different kinds of power, both of which are compatible with the view that power is something which is actualized rather than given. Mann's work incorporates both kinds of power, but the distinction between them is clearest in the work of Parsons and, following him, Giddens. On one hand, we have spoken about power as a means to get things done, a facility to secure certain goals and, on the other, as a means of constraint, a 'leverage' over others. The difference between them is perhaps best understood by contrasting the 'power to' side of things with that of the 'power over' dimension. Whereas the former rests upon enablement and tends to stress the possibility of collective, integrative action, the latter term refers to an instrumental ability to gain at the expense of another. Where one sees the possibility of collaboration, the other sees the potential for domination.

As noted earlier, Gidden rebukes Parsons for playing down the conflict of interests in society and attempts to maintain a balance between power as a transformative capacity and as domination. Parsons is only too aware, however, that the attainment of collective goals at the level of the system as a whole requires an organizational means to secure them. Organizational leadership is said to require the development and distribution of power in an unequal manner as a prerequisite for collective goals. Thus the institutionalization of control and co-ordination, in itself a form of 'power over' others, is for Parsons a necessary feature of the collective political landscape. Mann follows Parsons directly in embracing this line of thought. For him too, the collective and distributional aspects of power are entwined. The achievement of common goals may only be realized through the use of 'leveraged' organizational means. But to this, Mann adds a further conclusion.

Collective power, although integrative in spirit, may be used to further the sectional interests of like-minded groups or institutions to the detriment of other, less collectively organized individuals and bodies. There is no legitimate common good, only joint powers acting to the disadvantage of others. This, then, is an instance of 'power over', but one organized through collective association. Such practices are evident, for example, in the tie-ups in the global semiconductor industry, where the major players, the likes of IBM, Toshiba and Siemens, have been known to pool their resources to develop and exploit the next generation of semiconductors. At risk in such a scenario are the firms and workforces of different countries that may lose out in the development race. In Mann's terminology, such firms would have been 'organizationally outflanked' (1986: 7). A similar outcome could be deduced, as indicated earlier, from the collective powers of the middle classes to exclude the less organized from their neighbourhood preserves. Instead of integration and empowerment, qualities frequently associated with collaborative endeavours, we are back with division and domination. The 'power to' achieve is made to serve the power held over others. The circle is closed in an altogether familiar way (see Allen 1999a, 2002).

Not all accounts of the powers of association lead to this conclusion, however. An exception is the work of Hannah Arendt, whose view of collective action stresses rather than subsumes the collaborative, enabling side to power.

Arendt on collective power

Power for Arendt is a benign rather than a repressive force which is produced by people coming together to pursue a common purpose.

In this, she shares with Parsons the pursuit of collective ends, but unlike Parsons she does not view their achievement through the efforts of an institutionalized political leadership based upon binding obligations and sanctions. For Arendt, power is not generated 'from above' by a mobilizing leadership, but solely through association. It is rooted in *mutual action*:[4]

> *Power* corresponds to the human ability not just to act but to act in concert. Power is never the property of an individual; it belongs to a group and remains in existence only so long as the group keeps together. When we say of somebody that he is in 'in power' we actually refer to his being empowered by a certain number of people to act in their name. The moment the group, from which the power originated to begin with disappears, 'his power' also vanishes. (1970: 44)

In what is arguably her major political work, *The Human Condition* (1958), she goes on to show how power is a rather tenuous production, springing up between people only when they come together in action and debate around common, agreed, 'public' concerns. In this sense, power not only comes into being through association; it is something which actually holds people together in the pursuit of their common, agreed ends. The moment that a coalition of interests falls apart or the moment that collective action taken to achieve agreed ends dissolves, power itself is said to vanish. For Arendt, then, power is not something which can be applied like an instrumental force or measured or predicted like military strength; it is always of the moment. When it is not sustained by mutual action, it passes away. Or in her own words, power 'can only be actualized but never fully materialized' (1958: 200).

In adopting what must be the antithesis of the view that power is a thing, capable of being possessed, she is also able to demonstrate why it is largely independent of material factors such as the size and volume of resources in play. Popular revolt against a vastly superior force, or the ability of smaller players to mobilize effectively against the demands of those greater in size are possible precisely because of the collaborative, integrative resources

---

4 Jürgen Habermas (1977) was drawn to Arendt's conception of power primarily for the stress that she placed upon the formation of a common will through unrestrained communication, or rather speaking and acting together in civic spaces. He found fault in her understanding of power, however, for its Aristotelian roots, which located action in a public realm that bore no relation to what he took to be the systematically restrained nature of communication in the contemporary political arena. See Howell (1993), though, for an interesting discussion on the difference between Arendt's grounded notion of public space and Habermas's more abstract, formalized sense of the public sphere.

mobilized. Such qualitative resources are regarded by Arendt as every bit as significant as those that rely upon numerical strength for their edge. Moreover, the mobilization of such qualities is viewed as a positive gesture, as empowering of those taking part, rather than simply an act of resistance directed towards a more powerful, dominant force.

Clearly there are strong echoes here of the 'power to' side of thought, but what is underlined and indeed what has attracted the attention of feminist thinkers in particular is the positive, enabling conception of power upon which it is based.[5] Association, in this context, is not considered merely as a form of resistance or defiance against those who 'hold' power or as a collective endeavour aimed at bending another's will. Rather, the aim is the formation of a common will which enriches public life, and in so doing reaps benefits for all involved, especially in the way that people relate to one another. At the root of this conception, as may now be apparent, is a concern with the quality of *public space*, and in particular the moral, ethical and political concerns which raise rather than diminish the civic community.

Arendt's account of public space has been the subject of much critical debate, not least for its grounding in dubious ideals which reach back to the classical times of the Greek polis. Seyla Benhabib among others, however, has sought to interpret Arendt's views on public space in a manner that clearly reveals the positive powers of association. For Benhabib:

> public space emerges whenever and wherever, in Arendt's words, 'men act together in concert'. On this model, public space is the space 'where freedom can appear'. It is not a space in any topographical or institutional sense: a town hall or a city square where people do not 'act in concert' is not a public space in this Arendtian sense. But a private dining room in which people gather to hear a *Samizdat* or in which dissidents meet with foreigners become public spaces; just as a field or a forest can also become public space if they are the object and the location of an 'action in concert', of a demonstration to stop the construction of a highway or a military air-base, for example. These diverse topographical locations become public spaces in that they become the 'sites' of power, of common action co-ordinated through speech and persuasion. (1992: 93)

---

5 Seyla Benhabib (1992, 1996), Nancy Hartsock (1985) and Lisa Jane Disch (1994) have sought to show how Arendt's views can be understood as a feminist alternative to competitive, divisive accounts of power formation. Hartsock, in particular, is enthusiastic about the rejection of the notion of power as domination over others, and keen to show how power as mutual action is in line with much feminist thinking on power as competence and enablement rather than dominance. In a related sense, Disch wishes to argue how Arendt's notion of power rests upon plurality as a condition of public space and is in keeping with Iris Marion Young's recognition of a 'heterogeneous public' (Young 1990).

'Sites' of collective mobilization which draw upon expressive resources to make their voices heard are, on this interpretation, not directed towards the pursuit of sectional interests. Rather, such forms of social solidarity are intended to transcend particular interests by mobilizing around issues which are faced in common. The associational politics often identified with human rights issues is a case in point. The aim of mobilizing against human rights abuses and violations is not to replace one dominant group with another, previously subordinate group. In seeking to bring about a universal political and legal framework for promoting individual rights, one which is not influenced by the internal affairs of nation states, human rights activists such as Amnesty International argue that the eradication of all human rights abusers is a problem held in common and would be of benefit to everyone. In the construction of such a broad alliance, people of all different nationalities, genders and races would be empowered in ways which would be seen to integrate rather than divide them.

What is also interesting about this example, or for that matter campaigns such as the relative successes of the Jubilee 2000 debt relief campaign and of the International Campaign to Ban Landmines, is that the collective act of mobilization is itself regarded as a resource, on a par with any of those mentioned earlier, such as finance or property. The activation of moral and political energies, to draw upon Arendt's lexicon, and the ways in which they are channelled to influence and appeal to wider audiences – some of whom may be resource-rich in the traditional sense – illustrate the powers of association nicely. More than that, where such networks of alliance are organized across national borders, such energies may be directly supportive and enabling for groups seeking to bring about change in their own countries. Global networks campaigning around women's rights, say, on abortion, development and debt relief, human rights, or any number of environmental issues represent a form of social solidarity which transcends local, community notions of public space and can effectively turn such 'sites' into distanciated networks. Although Arendt appears to assume close proximity as a pre-condition of association, the mobilization of such energies 'at a distance', often orchestrated through the internet, is perfectly feasible across today global networks.

Indeed, if we think back to Giddens's line of argument that the 'stretching' of social relations is actually generative of power, then the extension of such networks of association may arguably be seen to enhance the amounts of power in circulation. The powers of certain environmental groups, for example, and their means to mobilize on a world-wide basis are enhanced both through their ability to represent environmental degradation as a 'global' problem which faces everyone and also through their ability to campaign across different 'public spaces'. The most effective environmental groups for example, such as Greenpeace and Friends of the Earth, campaign around issues which they claim to be of ecological significance not just to individual countries but to the planet as a whole. As such, their appeal to the citizens of the planet is intended to forge a set of common interests around the environment which, through cross-national agreement and co-ordination, enhances the persuasive power of these organizations. In fact, the distanciated nature of many environmental issues, such as those of acid rain or global warming, where the production of the effect and its harmful consequences do not coincide in space or time, may themselves provide the possibility for such a networked mobilization.

Moreover, despite the fact that such campaigns may be directed towards the cessation of particular economic activities which lie behind something like acid rain, and by implication specific economic interests, the collective endeavours involved may be understood as having less to do with the notion of winners and losers and rather more to do with a positive-sum scenario from which all may benefit. With the aim of enlivening public debate over the deteriorating state of the global environment, it could be argued that the various international networks of environmental groups have attempted to construct a public space for the deliberation of alternative, more consciously aware ways of living, as well as to promote reflection on different ways of organizing political and civic life. As such, some coalitions are likely to come close to Arendt's ideal of public spaces as an arena for action and informed debate.

From all this talk of campaigns and single-issue concerns, it is possible to gain the impression that the appearance and disappearance of power lend themselves to a rather short-term, unpredictable view of politics. People come together across varying distances united around a mutually agreed purpose or goal which, once realized or displaced, dissolves one public space in favour of another, more pressing goal. However, as Benhabib has pointed out, the agenda-setting nature of Arendt's associational power is not the central issue of political life: what really matters is the way that political life and dialogue are conducted. For Arendt:

> the effect of collective action in concert will be to put ever new and unexpected items on the agenda of public debate. Thus the 'associational' model develops not a substantive but a *procedural* concept of public space which is in fact compatible with this view. What is important here is not so much *what* public discourse is about as the *way* in which this discourse takes place: force and violence destroy the specificity of public discourse, by introducing the 'dumb' language of physical superiority and constraint and by silencing the voice of persuasion and conviction. Only power is generated by public discourse and is sustained by it. (1993: 105)

So the image of the fleeting moment which temporarily unites people around a common cause is less significant than the manner in which issues are debated and acted upon. Violence, above all, limits the possibility of people coming together to pool their collective energies in the pursuit of common goals. Acts of violence, the forceful suppression of others, are for Arendt the antithesis of power: they are a one-sided, instrumental act which invariably 'comes into play when power is being lost' (1970: 53) or is in jeopardy. The French government's violent action against one of Greenpeace's mobilizing campaign ships, the *Rainbow Warrior*, in New Zealand in 1985 forcefully illustrates the point, as does the violent repression of China's embryonic democracy movement in Tiananmen Square, Beijing, in 1989 by the central Chinese state. In the latter case, in particular, the separation and isolation of activists, the denial of association and communication around issues of democracy, robbed those involved of their collective power.[6] 'Violence', Arendt argues, 'can always destroy power; out of the barrel of a gun grows the most effective command, resulting in the most instant and perfect obedience. What never can grow out of it is power' (1970: 53).

On this score, power and violence are not only seen as opposites, but are almost bound to appear as located in quite different types of collective organization. Indeed, there is much to support this observation in that as power, for Arendt, can only be kept alive through association, so violence is embodied in overarching institutions like the state. On closer inspection, however, Arendt does not contrast violence relying on command and repression organized 'from above' and power fleetingly organized 'from below'. Rather, the contrast is drawn between those organizations which act on the basis of mutual consent and agreed goals and those which do not. Governments, for instance, may effect a monopoly on the means of violence and use it as a last resort to maintain their power, but that does not make them tyrannical regimes. Government as an organized activity may still rest on legitimate consent despite its monopolization of the means of violence.

The contrast in organizational forms that Arendt is reaching for can perhaps best be illustrated by drawing attention to the utter contempt in which she held bureaucracy as a form of domination, or rather, as she termed it, 'rule by nobody' (1970: 38). If we think, for example, of Benhabib's observation that a town hall where people do not 'act in concert' falls short, in Arendt's view, of becoming a public space, then Weber's bureaucratic official operating a rule-bound system of control falls outside of the frame of power. Organizations which draw their authority from expert

---

6 Arendt's work on totalitarian and tyrannical rule is the guiding influence here: tyranny's arbitrary rule silences individuals and destroys relationships between people, whilst totalitarianism is more insidious in that it mobilizes people as atomized political subjects in its support. The Nazi regime and Stalinist Russia are the two examples of totalitarian rule explored by Arendt in *The Origins of Totalitarianism* (1951).

systems and clothe their actions and judgements in an anonymous set of rules and procedures typify the hierarchical notions of power which Arendt rejects. Public officials and private managers who operate on the basis of impersonal, bureaucratic reason define a form of governance as control *over* others which she refused to acknowledge as power. Instead, she invoked a vocabulary of power which does not rely upon the language of domination and control.

For want of a better description, Arendt's vocabulary of power is concerned with the *transverse* relationships of power; that is, those which cut across conventional organizational lines and practices. The networks of association put together by many of those involved in human rights or green politics are, we have seen, engaged in collective action where the traditional notions of vertical power need not apply, or rather they do not apply in their conventional sense. Negotiation and persuasion are not always understood as modes of power, but in referring to the 'power to' achieve outcomes they are indispensable modes, and so too is authority – although in this context with an altogether different organizational meaning from that understood by Weber.

Authority, as a mode of power, has for Arendt lost much of its original meaning. In her essay 'What is authority?' (in Arendt 1961), she drew attention to a longstanding interpretation of authority which has its roots in the classical Roman tradition. In such a context, authority was something that was recognized rather than simply claimed; it was something that was held *among* people, not over them. Far from being instrumental as a mode of action, authority was considered to be closer to 'mere advice, needing neither the form of command nor external coercion to make itself heard' (1961: 123). Seen in this light, authority involved both informed knowledge and respect based upon the recognition of someone as *an* authority. Those holding such a position would be known in the community and deferred to, even if their advice was not heeded. Giddens (1994) refers to such figures as 'guardians', people whose judgement could be trusted on the basis of their 'wisdom' or access to a specialized body of knowledge, in contrast to Weber's rational-legal sense of authority. The idea of a hierarchical authority based upon technical expertise or impersonal rules stands in sharp contrast, therefore, to this more *lateral* sense of authority in the social community.

As in the previous chapter then, it is possible to broaden our understanding of the different modes through which power operates. In going beyond domination and other modes of instrumental power, what Arendt's arguments are able to illuminate is that associational power may also work through particular collaborative modes. Even where the same terms, such as 'authority' and 'persuasion', are in play, they can have quite different meanings depending upon the manner in which power is conceptualized.

We will return to the significance of this point in part II, where modes of power are considered in greater depth. For the moment, though, we turn to the spatial limitations of those accounts which adopt a view of power as something produced through the mobilization of resources, collective or otherwise, enabling or not.

Mobilized Powers, Stretched Resources

Despite differences of interpretation as to what purpose or ends power should be put to, Arendt shares with both Giddens and Mann the notion that power is produced through networks of association, or more accurately through networks of social interaction. While it is the instantiation of power in social action which represents the core of this conception of power, it is the mobilization of resources through networks which sets out the terms of its spatial vocabulary. In contrast to a vocabulary which describes power as distributed or delegated from any number of centres, on this view power is something that comes about through the 'stretching' of resources over space and time. When power is not sustained by networked interaction, it passes away, and it is in this sense that the 'stretching' of social relations may be said to be generative of power. This is clearly the case for Giddens and Mann, who both in different guises adopt a distanciated view of power, and it is more or less implicit in Arendt's view of collective action. Even though Arendt's account of collective mobilization is largely restricted to 'public spaces', an understanding of such spaces as collaborative networks makes it relatively easy to conceive of them as operating either in close proximity or 'at a distance'.

Yet as a spatial vocabulary of power, the notion of distanciated networks, I would argue, remains distinctly superficial. In Mann's case, whilst no doubt the organizational means available to different sources of power are important to register, if only because they alert us to the limitations and consequences of 'stretching' power over space, the analysis frequently stops at just the point where the more pressing questions occur; in particular, the question of how distance and mobility affect not just the extent of power, but also its mode and constitution. As with Giddens, there is a strong sense in Mann's work of how power acts as a medium through which distant goals are achieved across the networks. How far this type of analysis can go beyond that of a 'centred' conception of power, however, where power is transmitted intact, is quite unclear.

Clearly, mobilized powers are not transmitted outwards from an identifiable centre in any straightforward manner, but none the less there is every impression that power generated in one part of a distanciated network is transmitted intact across it. The crucial limitation of this view is that with power conceived as an effect of 'stretched' interaction, networks readily appear as little more than *carriers* of resources mobilized at different sites and locations on the network. When power, or rather, more accurately, resources, is or are said to be 'stretched' in this way, it is as if a metaphor of physical energy flows has been borrowed uncritically to convey the circulation of power.

Networked flows of power?

Despite the richer spatial vocabulary of this 'networked' conception of power, with its sense of reach, scope and action 'at a distance', the flow of power through the networks does none the less appear remarkably routine. The metaphor of flow is highly appropriate in this context, as it conveys the idea that getting things done 'at a distance', although mediated institutionally, occurs through the fluid transmission of resources. It is not as if little happens across the networks; on the contrary, the extent of mobilization at different sites across the different networks suggests considerable intervention. But, as in the previous chapter, power relations extended over space and time through the networks seem to remain relatively untouched by the experience. Power may be actualized by the mobilization of resources at particular points on the network, but neither the sites or power 'containers', nor the projection of power across space, are problematized.

Even in relatively complex accounts of networked interactions, such as that provided by Manuel Castells in his account of the rise of a global, information-based society, little attention is paid to the transformation – as opposed to the transmission – of power. In *The Rise of the Network Society* (1996), for example, Castells argues that inclusion and exclusion from networks, and the dynamics between networks, are the critical sources of domination which mark the beginning of an information age. Networks are conceived by Castells as open, interconnected structures, capable of unlimited expansion, through which resources of all kinds are said to flow: capital, information, technology, people, as well as images, sounds and symbols. Different kinds of networks are seen to involve different resources and each network is said to construct its own geography of control, with specific sites on the network producing, switching, directing and co-ordinating resources. By 'flows' Castells understands:

> purposeful, repetitive, programmable sequences of exchange and interaction between physically disjointed positions held by social actors in the economic, political, and symbolic structures of society... The space of flows is not placeless, although its structural logic is. It is based on an electronic network,

but this network links up specific places, with well-defined social, cultural, physical and functional characteristics. Some places are exchangers, communication hubs playing the role of co-ordination for the smooth interaction of all the elements integrated into the networks. Other places are nodes of the network, that is the location of strategically important functions that build a series of locality-based activities and organizations around a key function in the network. Location in the node links up the locality with the whole network. Both nodes and hubs are hierarchically organized according to their relative weight in the network. (1996: 412–13)

Whilst there is much that is moving around the networks in all manner of directions, therefore, as Castells so succinctly expresses it, it is 'the power of flows [which] takes precedence over the flows of power' (1996: 469). As power is transmitted through the networks, it is subject to switching and interruption in much the same way as energy flows are subject to redirection and resistance. Interactions across the networks may be constitutive of power, but in between the production of power and its transmission little other than directional changes across the hierarchies come into focus. Interestingly, the familiar example of 'world' or 'global' cities is offered by Castells as representative of a programmed network, where the decision-making powers of a New York or a Tokyo are of secondary importance to the constitutive power of the network itself. In place of an effective capacity for control, the global cities – as 'production sites' of the international economy – appear to draw their power from being locked into a networked series of programmable exchanges. As such, it is the power of the networked flows, the manner in which they enable cultural and economic elites to interact in simultaneous time, which draws Castells's attention.

Various public and private electronic networks, including the internet and company intranets, allow for continuous interaction to take place at the global scale in real time, which, whilst obviously facilitating the cluster of expert knowledges in particular cities, also brings together absent others in simultaneous time. The networks themselves appear to have a life of their own, the logic of which has the ability to carve out new spaces, 'very much like railways defined "economic regions" and "national markets" in the industrial economy' (1996: 412). In control of this logic, through their ability to direct and redirect the flow of resources, are the transnational elites, those who are based in the world's major cities, yet isolated from the majority of those around them. 'Elites are cosmopolitan, people are local' (1996: 415), states Castells as a means of conveying the point that such elites are attached more to the networks than to the cities themselves. What enables such elites to gain access to the networks and to circulate freely are the shared cultural and symbolic codes embedded in the flow of interaction, which serve both as a badge of admission and a barrier to entry.

Hence for Castells this is not an entrenched:

> 'power elite' *a la* Wright Mills. On the contrary, real social domination stems from the fact that cultural codes are embedded in the social structure in such a way that the possession of these codes opens the access to the power structure without the elite needing to conspire to bar access to the networks. (1996: 416)

Elite groupings in cities such as New York and London are seen therefore to mobilize their power through the networks, through the co-present interaction which enables them to shrink the space and time between them and to construct closer, integrated ties and connections. Thus the elites and the institutions of which they are a part are ultimately dependent on the flows and networks for their cultural and economic abilities, and are less influenced by, indeed are detached from, their local surroundings and context. The decision-making process is thus not located at all; rather it is actualized through the networked flows of interaction between cities.

Networks, we are told, constitute the new social morphology of our societies, and such:

> network morphology is also a source of dramatic reorganization of power relationships. Switches connecting the networks (for example, financial flows taking control of media empires that influence political processes) are the privileged instruments of power. Thus the switchers are the power holders. Since networks are multiple, the interoperating codes and switches between networks become the fundamental sources in shaping, guiding and misguiding societies. (1996: 471)

Yet amidst all this morphism the revisable and redefinable nature of the powers involved remains strictly out of focus, or more to the point untheorized. For all the detailed recognition accorded by Castells to the transmission of power through the networks, the absence of any serious consideration of power beyond how it is 'stretched', co-ordinated or redirected amounts to a rather limited conception of mobilized powers. Although the notion of power as a centralized force is passed over in favour of a more fluid conception, the unproblematized nature of power within and across the networks remains.

Moreover, this is the case whether it is a form of 'leveraged' or of collaborative power which is under consideration. Take, for instance, the case of overseas Chinese networks, an example which Castells also draws upon. Collaborative power, of the informal kind exhibited in overseas Chinese networks, is often portrayed as something which flows through the diffuse communities, be it Hong Kong or Taiwan or further afield in South East Asia or the large Chinese communities in North America. The networks of connections which links these communities are based upon family and regional ethnic ties, and sustained by particularistic ties of trust and obligation which enable things to get done 'at a distance'. The informal nature of these connections is often rooted in family chains that go back generations, with both personal and financial resources circulating through the dispersed communities, effectively bypassing national government controls to reach out across the globe (see Redding 1990; Olds 1995; Hsing 1998; Ong and Nonini 1997). That the 'power to' mobilize in such networks is frequently used at the expense of other sets of interests is beside the point here. As a form of collective association, the stability of these networks is assumed to be drawn from the pooled resources of loyalty and obligation. It is these qualities which, in Castells's view too, underpin the power of these networks to secure advantages along their length and breadth.

And yet, in considering these distanciated networks, there seems to be little appreciation that what actually holds them together over space and time amounts to more than the particular resources pooled.

Network Fictions

I started this chapter by outlining the sense in which power may be thought about as a medium, that is, as an effect produced in and through social interaction, and then, following Giddens and Mann, proceeded to draw a clear distinction between power and resources. In spite of what I then later said about distanciated networks amounting to little more than carriers of resources mobilized at different sites and locations, I do want to hold on to the importance of the distinction, especially for recognizing the critical role that resources play in the generation of power as an effect. Mann's insight into the logistics of (re)sources, in particular how they are organized through extensive and intensive networks, is a significant contribution to our understanding of what is involved in the process of resource mobilization. What Mann and others are also responsible for, however, is clouding the whole issue by creating the impression that power as well as resources actually flows or circulates over tracts of space and time.

The image of resources 'stretched' over space and time, organized in loose or tightly orchestrated ways to achieve maximum reach, is fine. So too is the idea of resources moving around different kinds of networks, controlled and co-ordinated from various locations, as part of an overall process of networked production. But it is a fiction to suggest that power flows in analogous ways. Power, as argued in the last chapter, is not some 'thing' or attribute; it cannot be possessed as resources can; and it does not travel.

How such an impression has been able to take such a firm hold in the imagination, enabling the likes of Castells to deliberate whether it is the power of flows or the flow of power which takes precedence, owes much, I would argue, to a spatial vocabulary of power that exudes plausibility. The analogy with physical energy flows is a familiar one, for the notion that power flows, or that it is switched, redirected or transmitted across the length and breadth of the networks, is likely to turn few heads. Castells's language of network hubs or programme exchangers serves only to augment the impression that social power, like its physical counterpart, circulates from point to point, from place to place.

Oddly, it seems that a particular view of social space which emphasizes the circulation and flow of power makes it difficult to consider it other than as bound up with problems of how to control and co-ordinate resources across space. Whereas networks are considered on this view to be straightforwardly constitutive of power in general, such accounts seem unable to entertain the view that spatiality may actually be constitutive of the various modalities of power called into play whenever and wherever power is produced and mobilized. Perhaps there is a parallel to be drawn between the idea of power as a fixed capacity, elaborated in the previous chapter, and power as a fluid transmission, outlined here. For both ideas serve to mislead, the former as a euphemism for resources and abilities, and the latter as a network fiction conveying a false sense in which power is believed to flow.

Having said that, I still wish to retain the view that networked relations of power are a significant part of our social architecture. It is the metaphor of power flows that I reject, not that of social networks. For some networked relationships of power are precisely of the kind that Giddens and Mann allude to, which allow for interaction to take place in near instantaneous time, whilst others involve a succession of practices bridging the gap between here and there in a more mediated fashion. A topological rather than a scalar conception of networks is to my mind the more appropriate means of expression, where power relations are constituted through their spacing and timing, whether they be instrumental in kind or of a more collaborative nature. To grasp this conception fully, however, where power is actively constituted in space and time, we need first to consider power as an immanent not an external force.

4

Power as an Immanent Affair: Foucault and Deleuze's Topological Detail

The sense in which power may be considered as an immanent force takes us far beyond the more accustomed ways in which power is thought to be related to the amount of resources at someone's disposal or to be possessed by an institution or a group. As it is commonly understood, power makes its presence felt, whether through direct action or through the collective efforts of individuals intent on pursuing a particular goal. Even if the power of, say, a software corporation or a government agency is deemed to be latent, it will eventually show itself in some manifest form. When power is *immanent*, however, this is not quite how it works. In this chapter, the aim is to show how it is possible to convey power as something which works on subjects as well as through them – at one and the same time. On this understanding, there is no external reference point to power, no force imposing itself from the outside to consider, only the sets of relations and circumstances that one finds oneself within. Power does not show itself because it is implicated in all that we are and all that we inhabit.

Perhaps another way to express this rather nebulous claim is to draw a simple contrast: if more familiar, 'centred' accounts of power consider it as something separate from what it can do, as a capacity distinct from its exercise, an immanent perspective conceives of power as *inseparable* from its effects. Power in this context has less to do with the extent to which people conform or comply with, say, the pronouncements of some authority than with the effectiveness with which subjects internalize their meaning. It is of little importance, then, to ask the question, 'Who has power?' and of greater concern to know something about the techniques of power which work as some kind of normalizing force. This understanding is undoubtedly clearest in the writings of Michel Foucault, where power is said to work through indirect techniques of self-regulation which make it difficult for individuals to constitute themselves in any other way.

For Foucault, and indeed for writers who adopt a similar view of power, such as Gilles Deleuze, the focus of this understanding is upon the techniques and practices which compose the texture of daily life. Whether institutional in form, as in the case of an administrative office or workplace, or wider in scope, connected to the welfare of subjects more broadly, the focus of inquiry is squarely on the endless play of techniques and practices which work to secure particular forms of conduct or, more pointedly, through which people freely fashion their own sense of self.

The stress upon the openness of outcomes, the fact that subjects are essentially *free* to regulate their own behaviours though not in all possible ways of choosing, is a key element of this approach to power. For one thing, it is suggestive of more than the idea that power enables subjects to get things done. Rather, it is suggestive of the open-ended nature of power, where even the best-planned techniques of power somehow do not take hold in the way imagined. The result, to borrow a phrase from Deleuze and Guattari (1988), is a kind of 'molecular soup', where unexpected elements come into play and things never quite work out in the manner anticipated.

It would be a mistake, however, to take this line of thought too far. Despite the uncertainty attached to this scenario, power as an immanent, normalizing force which operates through the detailed fabric of people's lives errs on the side of stability and regularity, rather than the reverse. Significantly, in an account which argues that power is not so much above us as around and among us, there is assumed to be limited room for manoeuvre. The emphasis is upon the closing down of possibilities, not the proliferation of them, and it is this attention to constitutive detail which provides the touchstone for a particular spatial vocabulary of power. If nothing is external to power, then the various techniques which place subjects, locate them or attempt to prescribe their movement in some fashion are integral to its very performance. In this sense, spatial assemblages, or rather specific ways of arranging social relationships, are part and parcel of the internalizing effects of power. They are, to be precise, constitutive of it.

This reformulation of power, with its stress upon the immanent nature of techniques and practices, is the main focus of this chapter. At the outset, we explore this formulation of power through the topological rendering of it by Foucault, and the related one by Deleuze, by concentrating upon its suggested diagrammatic qualities.

Whilst much of this analysis has taken place at the level of the institution, be it a site of medical practice, psychiatry, punishment or otherwise, more recently attention has turned to what it means to govern the self across a more diffuse landscape where subjects are less confined. This shift in the scope of analysis to the level of the entire population has, in my view, proved deeply problematical. In what little Foucault, in particular, left us to go on, the sense in which the exercise of government over a scattered population remains immanent in its dispersal is open to doubt. Lost in the topological detail, especially in the work of those who have attempted to adapt such a framework, I argue, are both what difference spatiality makes to power's effective reach and the particularity of the power relations involved.

As in the two previous chapters, though, before we proceed we need to be certain about what power conceived as a series of techniques, practices and diverse strategies actually entails.

Power as a Technique

Surprisingly, perhaps, the idea of power as a series of complex and diverse techniques is relatively straightforward to comprehend. In line with Foucault's (1982, 1984, 1988a, 1988b, 1988c) use of the term, there is no enduring capacity to power which may or may not be realized, only the routine deployment of techniques – spatial, organizational, classificatory, representational, ethical or otherwise, depending upon the forms of power involved – which seek to mould the conduct of specific groups or individuals and, above all, limit their possible range of actions. Techniques of power, in this sense, operate by inducing in others appropriate forms of conduct.

All of this, then, appears to be quite unremarkable – until, that is, one recognizes that the techniques of power only show up as an effect on the actions of others. There are no direct constraints on behaviour, no overt sanctions or prohibitions on what should and should not be done. There are only *indirect* techniques of regulation which work on self-experience. Commenting upon this aspect of Foucault's analysis of power in relation to sexuality, Charles Taylor (1985) notes that:

> this makes us objects of control in all sorts of ways which we barely understand. The important thing to grasp is that we are not controlled on the old model, through certain prohibitions being laid on us. We may think we are gaining some freedom when we throw off sexual prohibitions, but in fact we are dominated by certain images of what it is to be a full, healthy, fulfilled sexual being. And these images are in fact very powerful instruments of control. (1985: 162)

In this particular instance, the play of representational imagery would be one illustration of a technique of power deployed alongside others which make it all the more difficult not to assume a particular sexual nature. Techniques of the self, where people literally regulate their own behaviour and conduct, thus represent a novel aspect of Foucault's analysis, whether it be in relation to specific institutional behaviours in schools, hospitals, workplaces or prisons, or more broadly in relation to health, sexuality, the marketplace and so forth. In fact, the potential variety of arenas seems almost limitless and others, on the back of Foucault's work on the techniques of government in particular, have sought both to generalize the approach and to multiply the number of settings in which people make themselves up. Miller and Rose (1990), for example, in their approach to the government of self in economic life, treat us to a rich panoply of techniques and practices. Indeed, for them, the list *is* endless, as they sift:

> the actual mechanisms through which authorities of various sorts have sought to shape, normalize and instrumentalize the conduct, thought, decisions and aspirations of others in order to achieve the objectives they consider desirable. To understand modern forms of rule, we suggest requires an investigation not merely of grand political schema, or economic ambitions, nor even of general slogans such as 'state control' nationalization, the free market and the like, but of apparently humble and mundane mechanisms which appear to make it possible to govern: techniques of notation, computation and calculation; procedures of examination and assessment; the invention of devices such as surveys and presentational forms such as tables; the standardization of systems for training and the inculcation of habits; the imagination of professional specialisms and vocabularies; building design and architecture forms – the list is heterogeneous and is, in principle, unlimited. (1990: 8)

This may well be so. I find it difficult to believe, however, that almost anything and everything can count as a technique or relation of power. To allow ourselves to be guided by sound rules and conventions not of our own making is one thing. It is quite another, however, as Edward Said (1984) and others have pointed out, to say that they are always and everywhere a relation of power. Perhaps this highlights one of the difficulties of knowing what to include in an analysis of power when power itself is considered to be immanent in its dispersal.[1] This is where the idea of a topology of power gains in appeal, with its notion of specific *diagrams* of power.

---

1 It also prompts the further question of what falls within the remit of this chapter, given Foucault's widely influential account of power. A discussion of Said's work in *Orientalism* (1978) and *Culture and Imperialism* (1994), in particular his imaginative geographies of power which legitimate discursive constructions of dividing space between 'ours' and 'theirs', is probably the most significant omission. His central preoccupation with power and knowledge, in *Orientalism* for example, and the discursive representations which authorized certain views of the Orient, clearly speak to an immanent yet dispersed conception of power, albeit of a qualified nature. See, in particular, Gregory (1995) for a thorough examination and critique of Said's geographical imagin-

Diagrams of Power

There is something rather matter-of-fact about the notion of a diagram of power, in part because it lends itself to spatial imagery, but perhaps more so because it is suggestive of an easy mapping of power. Identify the points, intensities and distributions through the various techniques and strategies employed and, with a touch of cartographic zeal, it should be possible to sketch in the mass of capillary lines which connect the relations of force. The result, of course, would be excessively formalistic, but before such a mapping is dismissed out of hand there is, as Deleuze (1988) has argued in his examination of Foucault's ideas on power and subjectivity, considerable worth in thinking through the constitutive practices of power in diagrammatic form. The appeal for Deleuze of representing power diagrammatically – less than a metaphor for how power flows and rather more than an impressionistic piece of cartography — rests with its ability to set out the *interplay* of forces that constitute power.

What does this mean in practice? Well, expressed figuratively, Deleuze is keen to map the articulations and interrelationships between, say, the instrumental powers of a political or economical institution and the transformative powers of routine events which serve to recompose the lives of people in everyday settings. A diagram of power in this sense is said to be *coextensive* with its field of study, be it a firm's hierarchy spread across any number of regional sites, a diffuse network of political agencies, or an even more amorphous social body. The crucial element in all this is that this interleaving of relationships does not take place 'from above', directed from 'on high' as it were, but through the relationships produced *within* the various contexts and settings. If power works on and through subjects like this, rather than over them, then arguably it can only really be mapped as a spatio-temporal arrangement. As Deleuze (1988), commenting on Foucault's account of power, asserts:

---

ation. However, my main interest in this chapter is in what it means for subjects to govern themselves, and thus it is the *enabling* side to Foucault's grasp of power that I focus upon. For this reason, I also do not consider the recent Foucault-inspired work within political geography that concentrates upon the power/knowledge relation to explore the kinds of representational practices used by statespersons, elites and policy exponents to proclaim certain 'truths' about how and why political space is ordered, occupied and administered in the way that it is. This *geographing space*, the writing of it in ways that justifies authority over a subject population, has little if anything to say about how people 'work' on their own conduct to bring themselves into line. See, in particular, Ó Tuathail (1994, 1996), Ó Tuathail, Dalby and Routledge (1998), Dalby (1991), Dalby and Ó Tuathail (1998) and Dodds (1993, 1994, 2000).

The *diagram* is no longer an auditory or visual archive but a map, a cartography that is coextensive with the whole social field. It is an abstract machine. It is defined by its informal functions and matter and in terms of form makes no distinction between content and expression, a discursive formation and a non-discursive formation... If there are many diagrammatic functions and even matters, it is because every diagram is a spatio-temporal multiplicity. But it is also because there are as many diagrams as there are social fields in history. (1998: 34)

Or again:

The diagram or abstract machine is the map of relations between forces, a map of destiny, or intensity, which proceeds by primarily non-localizable relations and at every moment passes through every point, or rather in every relation from one point to another. (1998: 36)

Much of the inspiration for this graphic description of power relations comes from Foucault's early work, in particular in *Discipline and Punish* (1977), where disciplinary power is projected in diagrammatic form.[2] The diagram in question is actually of Bentham's infamous Panopticon, but Foucault's interest is in its *generalizable* quality: the practices and techniques of power which are mapped out by it. From the eighteenth century onwards, he argued, there has been a great deal of reflection on how the techniques of power embedded in this disciplinary diagram could be extended, adapted and put to wider use.

Uppermost among them in this context are the techniques through which the distribution of individuals leads to the imposition of certain forms of conduct through the spacing and timing of activities. The zoning, partitioning, enclosing and serialization of activities are critical to such a process. The arrangements of space, the particular assemblages of space which make up institutional complexes, are understood as integral to the ways in which particular forms of conduct are secured. In this line of argument, different spatial arrangements reflect the possible ways of acting inscribed in different schemas and serve to regulate, as well as enable, movement through them. As such, the disciplinary diagram, in an overt spatial sense, may be seen as having as much to do with fixing people down as with facilitating their distribution and circulation through an intricate web of relationships.

In *Discipline and Punish* this kind of diagrammatic mapping of the techniques of domination is not identified with any particular institutional

2 See also Foucault (1980), where the language of points, capillaries and net-like organization is deployed in relation to power, and the interview 'Space, knowledge and power' (2001b), where he conjures up an image of space as a map of relations between forces.

space, although the army barracks is recognized as an obvious candidate. Alongside the institution of the military, however, Foucault also draws attention to the disciplinary arrangements of schools, hospitals, factories, asylums and, of course, prisons. Stuart Elden (2001), in particular, has argued at length for such arrangements to be seen as part of a 'mapping of the present', where the spatial constitution of such institutional sites is fundamental to any exercise of power. Taking his cue from the meticulous, detailed survey of techniques that Foucault subsumed under the title of a 'new micro-physics' of power (1977: 139), Elden focuses upon the four acts of discipline outlined in *Discipline and Punish*, namely 'distribution within space; the control of activity-timetables, rhythms, dressage; the use of exercise; the articulation and combination of forces' (2001: 139), and takes the first to illustrate the mutual imbrication of space and power.

Alongside the raw facts of distribution and location, the layout, disposition and orientation of the various clinical or prison buildings, for example, are all deemed to have played a part in inducing particular forms of conduct, although not in isolation from the classificatory techniques and normative strategies designed to engage the minds of particular subjects. Significantly, the various techniques deployed were not simply used as a means of direct control over individuals; rather the aim was to enable not repress their subjectivity: to channel the conduct of individuals in a certain direction. Indeed, the target of such techniques and the medium through which they operate, in Foucault's own words, is the 'genealogy of the modern soul' (1977: 29).

The effectiveness of such calculated techniques for altering individual behaviour notwithstanding, there is clearly a logic and a rationale to them which in eighteenth- and nineteenth-century medical, educational, industrial, military and penal institutions makes it possible to discern a particular interplay of forces at work. The generalizability of many of the techniques from one institutional realm to another, the interconnections of powers both laterally and vertically, evoke a form of inducement and control which – in outline at least – invites comparison with certain contemporary forms of self-regulation (see especially du Gay 1996 on the inducement of particular forms of conduct in retail institutions). In particular, the stress placed upon detailed techniques of placement, the temporal elaboration of activities and their specific rhythms, as well as the specific distribution of individuals in relation to one another, lends itself to diagrammatic representation.

In making a case for the effectiveness of this technology of power, however, we need to be clear about how exactly Foucault and others consider it to operate. As with many commonplace notions of power and control, especially if we think of them as potentially mappable, there is always the risk, as noted earlier, that they become almost formulaic in style.

Once the points and distributions have been duly identified, all that remains to be done, as it were, is to join up the dots. This would then approximate to a form of cartographic determinism whereby the techniques in and of themselves are seen to induce certain forms of conduct, rather than those forms arising from the manner in which various techniques are interpreted, endowed with meaning, and practised. The latter is infinitely more scrambled and open to wider sets of possibilities than any 'inert' technology or diagram – as indeed Foucault was well aware.[3]

Commenting on the 'indisciplined' nature of institutional discipline, Foucault was quick to assert that:

> [t]hese programmes don't take effect in the institutions in an integral manner; they are simplified, or some are chosen and not others; and things never work out as planned. But what I wanted to show is that this difference is not one between the purity of the ideal and the disorderly impurity of the real, but that there are different strategies which are mutually opposed, composed and superposed so as to produce permanent and solid effects which can perfectly well be understood in terms of their rationality, even though they don't conform to the initial programming: this is what gives the resulting apparatus (dispositif) its solidity and suppleness. (1991: 80–1)

The key to the operation of power, therefore, lies with how the different schemas take hold in the imagination and serve to influence the timing and spacing of activities, rather than with any general formulae set down in advance. Thus the combination in a single diagram of discursive understanding and spatial disposition, channelled meanings and dispersed strategies, relies for its effectiveness on the extent to which people are absorbed by it. If people, whether in clinics or prisons, or a contemporary office setting or a public-sector housing estate for that matter, accept the 'truth' of the arrangements in which they find themselves, then those self-same arrangements hold real consequences for all those involved, providing a guide to what is and what is not deemed acceptable, when, and where.

On this view, what happens on a day-to-day basis at the office, on the estate, or even under a more obvious disciplinary regime experienced in a prison will not bear the formal imprint of an ideal programme, or reveal a

---

3 As too are Chris Philo (1992) and Felix Driver (1985, 1993) in their sensitive treatments of Foucault's 'substantive geographies'. In the kinds of spatial histories that Foucault elaborated in his 'micro-physical' studies, geographers were among the first to shy away from a formal geometric or dry cartographic interpretation of institutional spaces in favour of a more 'lived' interpretation of the 'disciplinary' arrangements. Perhaps a topological rather than a geometric interpretation, however, would have provided a better purchase on the precise 'geographies' involved in terms of how times and spaces are *actively* constructed in the relational arrangements of power. This topological understanding is developed more fully in part II of this book.

simple cartography of power. The everyday routines and rituals, as indicated, are more scrambled than that and yet, if Foucault is to be believed, it is precisely from this supple interplay of forces that power derives its reach and penetration. Elaborating on this point, he was keen to avoid the impression that something as pristine as the Benthamite technology owed its generalizable strengths to its classic architectural lines. On the contrary:

> these programmes induce a whole series of effects in the real (which isn't of course the same as saying that they take the place of the real): they crystallize into institutions, they inform individual behaviour, they act as grids for the perception and evaluation of things. It is absolutely true that criminals stubbornly resisted the new disciplinary mechanism in the prison; it is absolutely correct that the actual functioning of the prisons, in the inherited buildings where they were established and with the governors and guard who administered them, was a witches' brew compared to the beautiful Benthamite machine. But if the prisons were seen to have failed, if criminals were perceived as incorrigible, and a whole new criminal 'race' emerged into the field of vision of public opinion and 'justice', if the resistance of the prisoners and the pattern of recidivism took the forms we know they did, it's precisely because this type of programming didn't just remain a utopia in the heads of a few projectors. (1991: 81)

This gives us something of the flavour of Deleuze and Guattari's 'molecular soup' referred to earlier, where a kind of detailed topology of relations is in play in which the discursive and the spatial messages remain constant, yet the setting itself is in flux. On this account, regular forms of conduct are indeed induced, but not because they are 'read off' by subjects from a particular series of techniques or a particularly stark spatial arrangement. Rather it is the *interplay* of forces within a particular setting which makes it possible to extrapolate diagrams from the power relations inscribed within particular institutional spaces: subjects are progressively constituted, symbolically and practically, through specific points of purchase; mobilized and positioned through particular embedded practices; and channelled and directed by a series of grid-like expectations about how, when and where to conduct themselves and others. In simple terms, different kinds of diagram make different kinds of government and control possible, even through things rarely turn out quite as planned. Which, when put like this, is perhaps something that neither Michael Mann nor, say, the head of Microsoft corporation would find it that hard to agree with.

From this, we can now see the attraction of a diagrammatic representation of power for Deleuze. On the one hand, the systematic element to power relationships makes it possible to generalize about their (indirect) effects. On the other hand, the absence of any linear causal relationships to power whereby it is possible to trace simple outcomes make such generalizations open-ended. If the first point avoids an excessive nominalism, in which the analysis reduces itself to a mass of individual forces and their unique effects, the second alerts us to the task 'of analysing mixed forms, arrangements, what Foucault called apparatuses' (1995: 86). As Deleuze goes on to elaborate:

> We set out to follow and disentangle lines rather than work back to points: a cartography, involving microanalysis (what Foucault called the microphysics of power, and Guattari the micropolitics of desire). We looked for foci of unification, nodes of totalization, and processes of subjectification in arrangements, and they were always relative, they could always be dismantled in order to follow some restless line still further. We weren't looking for origins, even lost or deleted ones, but setting out to catch things where they were at work, in the middle: breaking things open, breaking words open. (1995: 86)

For Deleuze then, it is the open, yet differently articulated arrangements of power, the particular interplay of forces constituted in space and time, which form the basis of different diagrammatic representations of power. From this standpoint, it is but a short step for him to identify particular diagrams of power with particular sets of relationships and objectives: following Foucault's account in *The History of Sexuality* (1984), for instance, in relation to modern sexuality and the continuous regulatory channels and pathways of power that serve to qualify, measure and appraise 'useful' bodies, or in respect of the 'pastoral power' of the state and its techniques of confession (self-revelation), self-examination and guidance, used to draw out the individual and bring them into line with society's code of welfare and 'well-being'.

And, likewise, it becomes possible for others to identify broad programmatic relationships that lend themselves to a diagrammatic treatment, as in the case of Thomas Osborne and Nikolas Rose's (1999) attempt to show how different types of cities over time – from the nineteenth-century Victorian city through to the planned colonial and garden cities of the twentieth century, and its more contemporary 'enterprising' counterpart – may be understood as just so many diagrams of power. A diagrammatics of urbanisms, they claim, is made possible by attending to the immanent play of forces – their regularities as well as their distributions – which differentiate one type of city from another. This, they stress, is not simply about a collection of city plans, schemes or drawings, although the technical side should not be dismissed. It is also about how certain 'truths' about cities are generated to underpin particular kinds of urban existence and what it is to be in, say, a zoned city, where function and activities are separated in the name of health and welfare to foster a sense of the city as a living organism, replete with its healthy communities and 'normal' citizens. Urban subjects are in this sense mobilized, directed and channelled as much by an expectation of the persistent ungovernable aspects of a morally degenerate or diseased city as they are by the vision of a more virtuous urban existence.

In all such cases, whether it be the city that is diagrammed as a space of power or a society's code of welfare, power is conceived as immanent, yet dispersed: it constitutes its own organization, yet is produced relationally from point to point.

Oddly, though, this more expansive and diffuse topography sits rather awkwardly next to the meticulous and rather dense configurations of the prison or the clinic considered earlier, where particular institutional sites, not the population as a whole, were the units of analysis. The possible extrapolation of diagrammatic relations of power from the site of a prison, or from a factory, or even from the dispersed sites of a large company, is within the bounds of plausibility, but the idea of societal as opposed to institutional diagrams begins to push beyond that. Once the analysis moves beyond particular sites and specific institutions, the appeal of a diagrammatic representation of power, to my mind at least, starts to lessen as the lines, points and distributions take on more of an impressionistic form. For want of a better way of expressing it, the usefulness of thinking in terms of diagrams of power at the broad, 'macro-physical' scale, where the focus is the administration of whole populations, seems to me to be largely metaphorical. It is suggestive of an articulated schema of power, but little more.

Before I push this observation further, however, I want to step back and take a longer look at what exactly is involved when power is exercised in an immanent, normalizing fashion. For this, we need to turn to Foucault's account of government as 'the conduct of conduct'.

Governing the Self

In his seminal essay, 'The subject and power' (1982), Foucault outlines what may best be thought of as the indirect nature of power relationships. In contrast to a relationship of violence, for instance, which is direct, immediate and highly visible as a mode of action, for Foucault, the exercise of power shapes and limits the conduct of others behind their backs, so to speak. If not quite manipulation in the sense that the intent of power is concealed, there is, as indicated earlier, a degree of incitement or inducement involved. This can take any number of forms, ranging from the routine guidance of behaviour, through suggestion, prescription or authority, to more obvious forms of direction based on the control of movement and surveillance in its widest sense. Significantly, there is no place in this understanding for relations of command and obedience or for the more familiar expressions of force where one person imposes their will on another. There is, however, a closing down of possibilities involved which, as stressed by Lois McNay and others, involves a 'normalizing process of subjectification' (1994: 166):

> In modern society, the behaviour of individuals is regulated not through overt repression but through a set of standards and values associated with normality which are set into play by a network of ostensibly beneficent and scientific forms of knowledge. It is this notion of disciplinary power as a normalizing rather than a repressive force that lies at the base of Foucault's assertion that power is a positive phenomenon. (1994: 95)

So, on this view, power is something that works its way into people's lives through their acceptance of what it is to be or how they should act within particular contexts and scenarios. Indeed, the previous discussion of the significance of Bentham's panoptic technology revealed as much. If we push this a little further, however, what is also revealed in this understanding of power is the degree of complicity involved on the part of all those present. If somewhat less than complete incorporation, there is a degree of willingness to submit to the guidance or injunctions of others and a belief, held by those invoking the norm, in its beneficial outcome – that it is all to the common good. Power functions, then, through people 'working' on their own conduct, fashioning themselves in ways that reflect their acceptance of a particular norm which, perhaps unwittingly, makes them subject to its control.

For my part, Foucault's most illuminating illustration of this process is the technique of the confession. In the first volume of *The History of Sexuality* (1984), he sets out the nature of this ritual in which a knowing (authority) figure intervenes to guide or judge the actions of one who has a felt need to seek a better self. In order for the latter to reach a state of reconciliation with themselves, they have to first confront, reveal and then surmount the obstacles in their path before they are in a position to modify their conduct. Throughout the process, however, according to Foucault:

> the agency of domination does not reside in the one who speaks (for it is he who is constrained), but in the one who listens and says nothing; not in the one who knows and answers, but in the one who questions and is not supposed to know. And this discourse of truth finally takes effect, not in the one who receives it, but in the one from whom it is wrested. (1984: 62)

The subtle twists and turns of this scenario are instructive for a variety of reasons, not least because for Foucault the confessional form appears to capture many of the dynamics of the art of modern government. In the first place, it conveys the constitutive nature of power, where the relationship, the actions, the effects are all of a piece. There is no relation of force imposed from the outside, no external agency as in a scripted power play where one side advances its interests at the expense of another. Second, the revealing process highlights the productive, positive side to power which, in Ian Hacking's telling phrase, enables people to 'make up' themselves (1986: 234) through the process of self-reflection and self-regulation. People are free to govern their own behaviour in all kinds of ways, yet do so in ways that coalesce around the acceptable 'norm'. And third, the confessional form is founded on the presumption that both sides engage freely in the process, yet the knowing figure remains dominant throughout, guiding the conduct of the other and sanctioning their well-being (as in some professional–client relationships, for example). Perhaps more significantly, when this regulatory technique is associated with the art of government more generally it becomes, in Foucault's eyes, the nub of contemporary power: 'a set of actions upon other actions' (1982: 220) or in Colin Gordon's words 'the conduct of conduct' (1987: 296).

Intriguingly, when compared with the earlier discussion of disciplinary power and the likes of Bentham's Panopticon, this version of power seems even less instrumental in its execution and peculiarly aspatial in the play of its forces. It is possible, however, as Hindess (1996) in particular has argued, to see disciplinary techniques as one among a number of means of understanding how the conduct of others is influenced. So rather than posit a sharp separation between discipline and government, it is perhaps better to understand the latter as part of a broader, more supple strategy of normalization. If accounts of disciplinary forms of control have tended to err on the side of excessive domination, where individuals are left with less and less room to manoeuvre, as McNay (1994) and others have claimed, the shift in focus in Foucault's work to governmental forms of conduct represents an acknowledgement of the basic agency involved in power relationships. In place of disciplined subjects at the workplace, in factories or in institutions of confinement, the diffuse social and political arenas in which the art of government is exercised allow for the possibility of negotiation and even rejection of prescribed ways of acting and conforming. The *government of oneself*, for essentially that is what is involved here, thus entails a range of possible actions and reactions, albeit limited in their scope – otherwise the constraining element of power would be absent altogether.

In an oft-quoted passage, Foucault describes it thus:

> At the very heart of the power relationship, and constantly provoking it, are the recalcitrance of the will and the intransigence of freedom. Rather than speaking of an essential freedom, it would be better to speak of an 'agonism' – of a relationship which is at the same time reciprocal incitation and struggle;

less of a face-to-face confrontation which paralyzes both sides than a permanent provocation. (1982: 221–2)

Characterized in this way, power reduces itself to a series of provocations and incitements between individuals who still have options to play with. In agonistic relationships there is a certain kind of vying for position, a circling of one another as action follows reaction in an effort to influence the outcome of the process – not so much a scripted power play, then, as open-ended games of power within which the element of constraint has been whittled down to a minimum. Having said that, in true Weberian fashion the matter still comes down to how far which of the sides can successfully curtail or limit the choices of others. There is still an element which could be construed as domination involved in these relationships, even if it is not as clear-cut as those practised through disciplinary techniques of control. In fact, it is difficult to discern precisely what other modes of power are at work in the various ways in which individuals are said to constitute themselves as subjects.[4]

The issue here, which McNay (1994) has accurately identified, is that there are a great many ways in which the actions of individuals can be influenced by others, not all of which strictly speaking would fall under the heading of domination or for that matter could usefully be described as open-ended games. The inability to distinguish between the different ways in which power may be exercised or, worse, the inability to determine what is not a relation of power, in my view, seriously weakens Foucault's analytics. In chapter 2, attention was drawn to the shortcomings of treating power, domination and authority as synonymous and conflating the peculiarities of the different modalities of power. Working within Foucault's immanent conception of power, where power relationships have lost much of their specificity, there is a real danger that it becomes impossible to distinguish an act of manipulation from a gesture of authority, or an act

---

4 It is also difficult to discern precisely what significance should be attached to acts of resistance when power is portrayed as a series of provocations between different sides. Certainly the government of oneself as involving a series of actions and reactions could occasion resistance to something, although what explanatory value is gained from this labelling is not entirely clear. The 'power to' act which forms the basis of facilitative notions of power discussed in the previous chapter seems to me to allow for the idea of reactions which both enable and empower. Indeed, Hannah Arendt's associational sense of power is a testament to that possibility. For a non-oppositional account of resistance, see Pile (1997) and the response to it in the collection *Entanglements of Power: Geographies of Domination/Resistance* (2000) by Joanne Sharp et al., which argues that action and reaction run through *all* instances of domination and resistance. In what is otherwise a thoughtful account of the possibilities of resistance, where the term is opened up in all kinds of subtle ways, I feel it odd that the editors of the latter fail to do the same for power and merely equate it with domination.

of domination from a coercive threat, or the play of seduction from a form of social interaction such as persuasion.

The importance of establishing the specificity of power relationships and their distinct effects has been one of the main threads of argument running through this part of the book, not least to reintroduce some of the rigour lost in more diffuse conceptions of power, but also because I consider spatiality to be integral to the ways in which different modalities of power work themselves out in practice. Ironically, it is in accounts of the more dispersed art of government – in contrast to the more thoroughly centred diagrams of power – that power seems to have lost most of its distinctiveness.[5]

The art of dispersed government

The shift in Foucault's thinking away from a more constrained, diagrammatic sense of power as domination to a more open-ended series of provocations and incitements between individuals signals the shift in the focus of his investigations from the practices of institutions to those of government. This shift has been well documented by Colin Gordon (1987, 1991, 2001) among others, who, in making the case for the move from a 'micro' to a 'macro' analysis of power, emphasized Foucault's constant concern to document the detailed practices and specific rationalities of power, whether they be about the regularities of prison life or those extended over entire populations. The switch in focus, from the embedded institutional practices and knowledges through which individuals constitute themselves as subjects to the more dispersed technologies of normative constraint and benevolent guidance, according to Gordon, was Foucault's way of drawing together the different levels of analysis – without losing sight of the ways in which power is said to work its way through people's lives.

Foucault's lectures on 'Governmentality' in the late 1970s and his Tanner lectures delivered at Stanford University, in particular ' "Omnes et singulatim": toward a critique of political reason' (2001a), provide a

---

5 Mitchell Dean's (1999) *Governmentality: Power and Rule in Modern Society* and Nikolas Rose's (1999) *Powers of Freedom: Reframing Political Thought* both draw their nominalistic framework from Foucault's work on the art of government and both lose anything distinctive about power under a multitude of techniques and rationalities that are assumed to be heterogeneous and pervasive. While both texts are excellent at extrapolating ideas from Foucault's brief outline of governmentality and giving shape to what a contemporary analytics of government might look like, their casual conflation of power, domination and authority sits uneasily alongside the scrupulous attention that they pay to various technologies of power.

sketch of how the government of oneself is possible across what can only be termed a scattered and diverse population. Instead of the meticulously recorded detail of specialized practices in the fields of medicine or punishment, we are treated to a broader canvas of immanent techniques characterized by their disparate, yet pervasive nature. Whereas, conventionally, the sovereign powers of the state are portrayed as a force exercised narrowly on the basis of consent or extensive repression, government for Foucault is something that permeates the conduct of individuals across a wide spectrum of activities and senses of self: as a member of a household or family, as a worker, a citizen and so forth:

> In contrast to sovereignty, government has as its purpose not the act of government itself, but the welfare of the population, the improvement of its condition, the increase of its wealth, longevity, health, etc; and the means that the government uses to attain these ends are themselves all in some sense immanent to the population; it is the population itself on which the government will act either directly through large-scale campaigns, or indirectly through techniques that will make possible, without full awareness of the people, the stimulation of birth rates, the directing of the flow of populations into certain regions or activities, etc. (1991: 100)

Government, then, like discipline, is an immanent affair; it is assumed to reach deep into the lives of a disparate population to the extent that people more or less internalize its effects. The mix of direct and indirect techniques of regulation which mould our self-experience are said to make us, in Taylor's (1985) words, objects of control in ways which are barely understood. So how, then, is the management of a population possible in such depth and with such attention to detail? On what basis does power achieve such a penetrating reach?

The answer to these questions, as intimated earlier, lies with the *freedom* that people have to govern themselves. As Gordon (2001) observed, Foucault's exploration of the workings of government enabled him to address the relationship between power and freedom in a more revealing way than hitherto. The practice of government 'encapsulated the key insight that power, understood as a form of action on the action of others, only works where there is some freedom' (2001: xxviii). In the absence of any sanctions, we opt to restrain our behaviour because we may freely choose what is appropriate and what is inappropriate behaviour. Thus, in relation to family life, for example, we may possibly think that in challenging the authority of the father-figure in the family setting we gain a certain freedom, yet in Foucault's reasoning we are just as likely to remain constrained by 'truths' and practices associated with 'normal' family life and its relationships of parenting and appropriate sexual behaviour. In short, we remain 'trapped' in our thinking about the nature of 'proper' families, even though we may have set aside the shackles of patriarchal authority and the traditions of male order.

More generally, Foucault takes the freedom that people have to get things done or to make themselves up in certain ways as a necessary part of what it means to govern. A degree of freedom is implicit in the art of governing, in the liberal sense that the promotion of freedom, rather than its denial, is the most efficient way of achieving governmental ambitions. Excessive government, the direct intervention into the welfare and safety of citizen's lives, for example, or the visible imposition of constraints on personal conduct, turns out, on reflection, to be far more time-consuming and cumbersome as a means of bringing people into line than does provoking them – as free agents – to act in ways which they would otherwise not have chosen. This is the kind of economic liberalism that Weber understood well, where people find themselves falling into line with the dominant market interests simply by following their own rational self-interest. In a similar although not identical vein, Foucault recognized the significance of people 'working' on their own conduct, free to fashion themselves in all kinds of ways, yet invariably narrowing down the possibilities in line with acceptable behaviour.

Where the two analysts of power part company, however, in this instance is that for Foucault the forces in play operate *through* the constitution of various forms of subjectivity. The relation is one of immanence, not externality. The management of a population or society in depth is possible only on the basis of freely self-constituted subjects. The assumption is that such power, to follow Foucault, both individualizes *and* totalizes in the fullness of its scope.

This is, to say the least, an heroic assumption and one spelt out in general terms in his lectures on government as 'of all and of each' – *omnes et singulatim*. In the Tanner lectures we are offered a glimpse of what is involved when power is said to come from everywhere to shape and influence the welfare of subjects – both as individuals and at the level of the population as a whole. The dispersed technology in question is that of 'pastorship', a loose metaphor for describing the kinds of obligation and care that welfare authorities (read 'shepherd') have for their (flock of) people.

Rooted in Hebrew and Christian tradition, the notion of pastorship entails an obligation on the part of the authorities to guide (to watch over) and to unify (to gather together) what is, in effect, an aimless group of dispersed individuals. The guardian's obligation to manage economically and kindly, however, involves a reciprocal responsibility on the part of the individual to give themselves over to the will of the guardian. Although one is nominally free to choose one's own welfare, the risk of losing one's way implies a need for the individual to examine themselves in such a way that they are able to be more effectively guided. A knowledge of oneself, an ability for self-reflection and self-regulation, thus becomes a prerequisite for individuals in their search for a better, healthier and wealthier self. The technique of confession, noted earlier, where both sides engage freely in the process, yet the knowing figure remains dominant throughout, guiding and sanctioning the conduct of the other, is the model for this individualizing practice. It captures the sense in which people willingly place themselves in a subordinate position in the hope of securing a better life.

Set within a more up-to-date language and context, government bodies, trusts, voluntary organizations and any number of quasi-welfare agencies, from private medical and educational institutions through to the corporate care of employees (the welfare 'police' of old, to follow Foucault), work within a framework of expectations and presumptions that enable others to know what is expected of them and to adapt accordingly. Such bodies diffuse standards and exhort us to ideals which are hard to find fault with, and provoke us into taking responsibility for ourselves in ways that more or less mirror acceptable types of behaviour. It is in this sense that power individualizes, reaching deep into the self, it is claimed, because, in Foucault's words, 'it is coextensive and continuous with life' (1982: 214).

On this view, then, the art of government rests upon the continuous and relatively stable presence of a series of ideals, expectations, received 'truths', standards and frameworks which provoke individuals to govern their lives in quite particular ways. Power, it seems, has a remote yet immanent quality, capable of reaching into people's lives because they alone bring themselves to order. This is possible, even probable, but we are being asked to accept this outcome with little understanding of the spatial and temporal arrangements involved. In contrast to the detailed survey of techniques in Foucault's earlier institutional analyses – documenting the distribution of individuals in penal spaces, for example, on the basis of a series of grid-like expectations about how prisoners should conduct themselves – we have scant detail of the spatial assemblages involved in the management of dispersed populations.

One attempt to fill this topological gap is to be found in the work of Deleuze and Guattari, in particular their essay on micropolitics and segmentarity in *A Thousand Plateaus: Capitalism and Schizophrenia* (1988). For them, the dispersed arrangements of power described by Foucault are closer to the open-ended games of power that he outlined than to anything more cartographically certain. In an inventive vein, they consider the exercise of power to be something of a hit-and-miss affair, with attempts to fix a particular coding or framework of expectations limited in their ability to guide or constrain actions. Power, in their supple hands, is conceived as internal to what it can do, and what it can do is the outcome of its immersion in organizational forms where all manner of reinterpretation, fluid negotiation and translation hold sway. In this more uncertain script, the variety of institutions in which power is said to be 'centred' – state agencies, private corporations, financial institutions and religious establishments among them – aim at best to adapt and convert the fluid lines of social activity into more stabilized segments. Caught up in the intense flow of social activity, however, such 'centres' are conceived as unstable configurations quite unlike their more boxed counterparts discussed in chapter 2, open to mutation in altogether new and often unanticipated ways.

Now, we can make of this diagrammatic sensibility what we will, but the recovery of power as a more open-ended, hit-and-miss affair eases the pressure to account for the actual effects of so many dispersed techniques and practices of government. If we have to worry less about the detail and depth of power's reach, however, the question remains as to how the successive and cross-cutting practices of government are mediated through space and time. If government *is* an immanent affair, constituted in space and time, as Foucault has argued, how does power bridge the gap between here and there to govern a diffuse population?

Immanent Powers, Dispersed Technologies

One answer to the above question would be to revisit the view expressed earlier that power constitutes its own organization, yet is produced relationally from point to point. The stress here, you may recall, was not upon topology for its own sake, as a rather abstract configuration of points, but upon the interplay of forces within a particular setting, which made it possible to talk about pathways of power, regulatory channels, points of application and the like. The broad idea that power may be coextensive with its field of inquiry gives us a feel for the kind of inducements that are *placed* through a social body rather than handed down from a point above and beyond it. Subjects are constituted by the spacing and timing of their own activities as much as they are by those of others who seek to influence their behaviour; their conduct is shaped as much by what they absorb and imagine the 'truth' of their circumstances to be as it is by the physical layout, distribution and organization of their surroundings.

All of this, it seems to me, is insightful, and exceptional in its grasp of the roundaboutness of power, but unfortunately it has little to tell us about how a diffuse population is governed. In so far as the idea of a diagram of power at the 'macro-physical' scale takes us little beyond the realm of metaphor, so the embedded spatial assemblages of power within particular settings provide few clues as to how the government 'of all and of each' is achieved on a stable and continuous basis. If the aim of government (for Foucault does believe that power is intentional and non-subjective, 1984: 94) is really about the subjectification of a multitude of dispersed individuals, then no simple aggregation of site-specific techniques will deliver such an objective. Bridging the gap between here and there to bring a diffuse population within reach is singularly unlikely on the basis of a scaled-up version of confined arrangements – no matter how hit-and-miss the affair.

Deleuze, it would seem, and one would guess Foucault too, knew this, or at least neither attempted to make the case for a rescaling of power from the 'micro' to the 'macro' arena. Rather their adherence to a topological version of space required that neither of them do so. Deleuze, in particular, had a surfeit of language with which to describe the apparently elastic reach of power, as in the case of various institutional 'centres' attempting to adapt the fluid lines of social activity into more fixed, stable arrangements. Yet as somewhat precarious combinations of social activity, the basic institutions of government are themselves said to be caught up in the intense flow of social interaction, open to mutation in often quite unexpected ways, yet constantly striving to hold things down, to regularize zones of activity.

In his essay 'Many politics', in a collection of his writings, *Dialogues* (Deleuze and Parnet 1987), this diagrammatic sensibility is outlined to show how each of us, whether as part of a group or as an individual, is caught up in an array of cross-cutting lines of social activity which effectively weave our lives into some kind of dense fabric. The resulting texture, to continue the metaphor, is both supple and rigid, open to the detours and contingencies which shape our lives, yet locked down and constrained by the exactitudes of family life, work and welfare, for instance. The largely predictable set of social practices which make up the latter and the manner in which they are governed by normalizing expectations and moral presumptions all serve to limit our lives, yet they remain in flux, pulled by unforeseeable forces that upset the predictable and undermine what passes for moral governance. Both sets of forces are in play, 'within each other at the heart of the assemblage' (Deleuze and Parnet 1987: 132), and both place us, it seems, as subjects in spaces in which we act and are acted upon to reproduce our subjectivity anew.

In the same essay, Deleuze describes to us in equally general terms how even the most centralized of state authorities is not 'the master of its plans', as it too is enmeshed in circumstances both supple and rigid, extending its reach in ways that are predetermined yet volatile in outcome:

> The most centralized state is not at all the master of its plans, it is also an experimenter, it performs injections, it is unable to look into the future... It is along the different lines of complex assemblages that the powers that be carry out their experiments, but along them also arise experimenters of another kind, thwarting predictions, tracing out active lines of flight, looking for the

combination of these lines, increasing their speed or slowing it down, creating the plane of consistence fragment by fragment. (Deleuze and Parnet 1987: 145–6)

If this is merely to say that events and plans do not pan out in altogether expected or anticipated ways, when indeed, we might say, was it otherwise? But in a more generous spirit, I take Deleuze to be pointing up the dispersion of power as an immanent affair, in which the fine detail of the powers that be are immersed in the 'molecular soup that nourishes them, and makes their outlines waver' (1988: 225). As an evocation of power, it has considerable merit, but the overall spatial imagery of various sets of interacting lines of force, supple or not, still leaves much to be said about what brings people within reach. If the topology is a bold one, the spatial configuration, for all its worth, remains frustratingly abstract.

In comparison, Foucault's topological musings on the art of government, as we have seen, verged on the aspatial. Governmental forms of conduct appear to be self-sustaining in character, with a reach that, in intent at least, is both all encompassing and individualizing. The willingness of 'free subjects' to regulate their own conduct in ways not necessarily of their own choosing holds the key to how government is possible at points remote from their application. In fact, according to David Scott (1995), there is no equivalent to the rich spatial vocabulary of disciplinary power at the level of government. The habituation of mind and body that discipline imposes upon the lives of those in institutional settings does not, he argues, have a direct counterpart at the 'macro' level. 'Government', he asserts, 'does not regulate in this kind of detail' (1995: 203) and he goes on to show how Europe's colonizing practices by the nineteenth century operated through inducement and obligation to bring about new forms of colonial subjectivity which, to those both near and far, would be deemed 'improving' (see also Gregory 1998).

Gordon's (2001) account of government as a succession of practices animated and justified by a succession of different rationalities also eschewed the kind of detailed spatial probing that the questions asked of disciplinary power demanded. Again, the assumption that people are free to govern themselves, a freedom that underpins the possibility of indirect techniques of government inserting themselves into the lives of distant others, is what apparently dispenses with the equivalent spatial curiosity. Once the collective and individual life of a population becomes the explicit target of the practices of government, it is as if the transformation of power relations *across* space is of less fascination or interest than those transformed *in* space, on site as it were.

This geographically skewed topology, I suspect, is the product of a conception of government that holds tenaciously to the view that power is always and everywhere immanent, yet is dispersed at the point of application. Somewhat bizarrely, this view reminds me of the discussion of globalization and multi-level governance at the end of chapter 2, where it was claimed that a new territorial reordering had occurred in response to the dispersion of power across various levels and sites of authority. Rather than a simple top-down transmission of power, you may recall that on this account, there is said to have been a redistribution of powers between the different levels of governance – from regional and national state agencies to non-state agencies and so on – which brought a more distant population within reach.

Now, to avoid any misunderstanding, I should add that it is not part of my argument to suggest telling parallels between the two views of power; the fact that one version operates on a scalar basis with power conceived as a force imposing itself from the outside and the other is topological and immanent in its conception should dispel such an idea. What struck a chord, however, was the unproblematized projection of power *across* space common to both views (the same is true for the networked versions of power discussed in chapter 3, but in fact they involved the transmission of resources, not of power). Whilst 'centred' views of power have yet to recognize the problem, the challenge for those who hold that power has an immanent presence is to grasp how, in the context of a diffuse population composed of a multitude of wills, the subject and power remain mutually constitutive of one another in space and time.[6] Fortunately, this is a challenge that has been taken up, even if not quite met.

Deterritorialized power

Michael Hardt and Antonio Negri's *Empire* (2000) is a provocative adaptation of Foucault and Deleuze's thinking to the question of globalization and

---

6 The work of Nikolas Rose, especially his *Powers of Freedom: Reframing Political Thought* (1999), has addressed the issue of government at a distance, but as his argument relies heavily upon the ideas of actor network theory, in particular the thinking of Bruno Latour and Michel Callon, this body of work is examined in chapter 6. Another reason for considering Rose's ideas in part II is his more explicit account of authority as a modality of power capable of reaching a dispersed set of wills. Earlier in the chapter, reference was made to Miller and Rose (1990) and along with other writings (Rose and Miller 1992; Rose 1993, 1996a, 1996b) their contribution is assessed more fully in the context of what it means to govern across space. Mitchell Dean's *Governmentality: Power and Rule in Modern Society* (1999) is a further attempt to elaborate what, following Foucault, it means to govern others and ourselves in a wide variety of contexts, but as with Gordon there is no spatial curiosity as to how power may be mediated in space and time. See also Dean (1996a, 1996b, 1998).

sovereignty at the turn of the twentieth century. It is, without any risk of understatement, an ambitious book, with the explicit aim of setting down how a new global cartography of power, which they call 'Empire', is exercised in immanent fashion. Whereas Deleuze has been known to talk about the extension of capitalism to the whole social body through the overbearing institutions of the world market (see Deleuze and Parnet 1987; Deleuze 1988), Foucault's ideas of discipline and government have, to my knowledge, yet to encounter the politics of globalization in such a spirited manner. In *Empire*, this topological shift makes for interesting reading, as it is now the space of the global that is said to admit no external reference point, that leaves nothing outside the reach of this new form of imperial sovereignty.

Empire, it should be said, is not just another piece of shorthand for a US-style hegemony in a world system of states; rather it is intended to convey a decentred, deterritorialized apparatus of rule that has no transcendental centre of power, no historical equivalent to Washington or imperial Rome. Having said that, this apparatus of rule is far from ethereal; rather it is 'composed of a series of national and super-national organisms under a single logic of rule' (2000: xii): a sort of grand amalgam of the United States in coalition with any number of 'willing' states, transnational corporations, media conglomerates, supernational institutions, NGOs and the like. And it is this logic of rule that manifests itself in an immanent fashion, through a series of modulating networked relationships, diffusing moral, normative and institutional 'imperatives' on, we are told, a global scale.

The spatial topology of power involved is left rather vague, but we are given a clue by the authors when they liken the world market to the *diagram* of imperial power. Organized on the basis of free subjects, the operation of the world market, or rather world markets, constrains those very same subjects by directing and influencing the economic and monetary frameworks in play. It is in this analogous sense that the networked relationships of power are said to progressively bring everywhere within the ambit of rule. As on many occasions, despite a reluctance to comply and without a hint of obligation, people bring themselves into line with the interests of the markets because they have no choice but to do so (domination 'by virtue of a constellation of interests', as Max Weber described it – the 'objective circumstances' which make submission the only realistic option, as noted in chapter 2). So, in equivalent political terms, according to Hardt and Negri, the institutional processes of normalization, including the tenets of neoliberalism, reach so far into the lives and subjectivities of individuals that it is no longer possible to discern their points of application, or even begin to question their 'taken-for-grantedness'.

More to the point, in these open, extensive networks, power is not seen as something that is applied externally by the likes of a 'superpower' such as the United States; rather the networks themselves are constitutive of the very power that enables the United States and its allied governments and organizations to act. Everything is, as it were, 'bundled' together (in the sense of Microsoft 'bundling' software inside its operating systems), so that global trade, open markets, human rights, democracy, freedom and much more are inseparable elements of rule: if you 'buy' (or are immersed in) the logic, you have no choice but to take its disparate elements. They are part of a seamless logic – a 'smooth space' of rule – where the surfaces of everyday life 'are crisscrossed by so many fault lines that it only appears as a continuous, uniform space' (2000: 190).

This, if I am not mistaken, is the land of the all encompassing and individualizing rule where new subjectivities take their shape from the simple act of living. Every time, for example, we don a pair of Nike trainers, wear a Gap sweatshirt, drink a Starbucks coffee, consume a genetically modified meal, or take advantage of a privatized health or education service, we 'buy' into a lifestyle package that has, as part of its make-up, a certain conception of freedom and of democracy, whether we like it or not. This logic works in and through difference, in true liberal style, taking whatever identities are to hand, and refashioning them to different ends. The inducements are so familiar, the subjects so willing, that we take it upon ourselves to absorb this 'improving' (and approved) lifestyle without, it should be noted, the need to be 'walled in' by any institutional arrangement.

If this is so, then it is indeed the case that the habituation of mind and body that institutionalized discipline imposes has, as Scott argued, no direct counterpart in the daily lives of individuals in the world at large. The diagram of power in this context has little to do with power constituting itself relationally from point to point and more to do with the immanent production of subjectivity. As Hardt and Negri express this shift in one of their more Deleuzian moments:

> This is precisely where the most important qualitative leap must be recognized: from the disciplinary paradigm to the control paradigm of government. Rule is exercised directly over the movement of productive and cooperating subjectivities; institutions are formed and redefined continually according to the rhythm of these movements; and the topography of power no longer has to do primarily with spatial relations but is inscribed, rather, in the temporal displacement of subjectivities. Here we find once again the non-place of power that our analysis of sovereignty revealed. (2000: 319)

Government, then, has finally lost any spatial reference point. The non-place of power echoes Naomi Klein's (2001b) evident frustration that contemporary power in a global age is so everywhere that it seems nowhere. How do you challenge a political order that is so amorphous in style, so 'bundled' together, as to leave no room, or rather no space, for political alternatives to take hold?

For Hardt and Negri, the answer is a counter-Empire, so to speak, which has its roots within the confines of the new sovereign order. In challenging the diffuse politics of globalization, the emergence of an alternative, more inclusive order comes about not through any number of local struggles at the margins, but through people, the dispersed multitude, pitting one form of global sovereignty against another. Instead of people mobilizing to 'escape' the influence of this new apparatus of rule, Hardt and Negri stress the need to build an associational politics that confronts it head on and comes out the other side. In the absence of unitary centres of power, protest and struggle mirror the 'non-place' of power which, to follow Klein's (2001a, 2001b) lead, chooses targets which symbolize the new logic of rule. MacDonalds is an obvious choice, as is Nike or Shell Oil, but so too is the World Trade Organization, the IMF and Kyoto. Thus, on this understanding, the immanent nature of contemporary global power has not only brought forth a counter-political movement in its own diffuse image, but has also led to the identification of virtual 'locations' of government which, paradoxically, have given power a spatial definition.

In this immanent landscape of power, then, apart from its symbolic locations, we appear to have moved beyond the need to consider the dispersed character of government, given its essentially *unmediated* nature. There is no body of power to overreach itself, because the immanent production of subjectivity does not operate relationally on a point-by-point basis. It works on and through everyone and every individual, but without spatial reference.[7] Yet, for me, there remains a certain hollowness to this conception; an emptiness in the analysis precisely where the spatial and temporal constitution of government's mediated interactions should be. It is deeply worrying, to my mind, not to be curious about the spatiality

---

7   Hardt and Negri's conception of power, or rather what appears to be Negri's conception, rests upon a particular interpretation of Spinoza's distinction between *potestas* (loosely, the power to act) and *potentia* (loosely, the capabilities or sovereign apparatus of rule). The actualization of *potestas* in *The Savage Anomaly: The Power of Spinoza's Metaphysics and Politics* (1991) – translated into English by Hardt, incidentally – is what Negri is later to make the cornerstone of counter-Empire. It also, however, underpins his republican notion of Empire where an unmediated, immanent form of power makes it all the more possible for the multitude supposedly to mobilize against it, in contrast that is to the more formidable, mediated *potentia* of the state and its associated institutions. In chapter 5, the parallels between Foucault's notion of power (*pouvoir*), as something exercised rather than possessed, and the Latin *potestas* are explored more fully.

of empire, for it is from the topological detail that, I think, power's mediated relationality is to be gleaned.

Topological Detail

Perhaps the odd thing about this chapter, especially in a book intent on revealing the geographies of power's proximity and reach, is that spatiality as a focus faded from view the more we moved beyond particular sites to consider the fate of the dispersed multitude across the globe. Once outside the walls of the institution, so to speak, it was as if a concern for the detailed spacing and timing of activities, and how they induced and channelled particular patterns of behaviour, no longer had any real purchase on the more expansive matters at hand. Whilst I do think that the practice of government over a diffuse set of bodies does require a different sort of spatial probing, it is a mistake, in my view, to lose sight of what is common to any spatial analytics of power.

In the first place, Foucault was right to stress that power should be understood as coextensive with its field of operation, as an immanent force which constitutes its own organization rather than one which imposes itself from the outside. In this respect it matters little whether it is the practices of punishment in institutions of correction that hold our attention or the dispersed techniques of neoliberal government. Both require that we understand power relations as constituted through their spacing and timing and not, I should add, as some imagined capacity held 'in reserve' ready to impose its will on the prison or the civic population. Equally, we should not think of power as something which flows or circulates through such populations to hold them in check.

Deleuze was also right to draw attention to the open, yet differently articulated arrangements of power which draw their provisionality from the interplay of forces constituted in space and time. The diagrammatic representations of power that flow from this analysis have, predictably enough, tended to freeze and formalize such arrangements to their detriment, but the topological appreciation gained far outweighs any loss in understanding. For one thing, it diverts our attention away from the effects of, say, the size of state or business institutions and the geographical scale at which they operate, and towards the relational arrangement of which they are a part. A topology of power, in this sense, is concerned with its relational constitution, from one location or point to the next, be they institutionally close, far apart or remote from one another.

It is at this point, however, if you forgive the pun, that those seeking to explain the practices and rationalities of dispersed government should have become spatially curious. What Gordon passed off as a succession of practices might, a more inquisitive mind might have thought, have set them thinking about the *mediated* relationships involved for any kind of succession to take place, in real time or otherwise. It might also have forced them to puzzle out why certain practices are more effective at a distance, whilst others require close proximity to have any real impact. Power's immediate and effective presence is not a given, and it is this kind of spatial probing which sets the agenda for the remainder of the book.

Part II

Lost Geographies

In the first part of this book, I explored the work of various writers for whom space has become increasingly central to the understanding of the trappings of power. I drew attention to the different spatial vocabularies of power expressed, from the more centralized, linear forms of power inspired by Weber's writings on organization and bureaucracy to the thoughtful analyses of networked and distanciated forms of power outlined by Mann and Giddens, as well as to the more mobile and diffuse diagrams of power drawn by Deleuze and Foucault. In each case, the spatial aspects of power were to a greater or lesser extent documented as part of the exercise of power, yet as I see it none was sufficiently concerned, with the exception of Foucault, to probe the difference that spatiality makes to the workings of power – where it is an integral, rather than an additional, part of the picture. As space is already implicated in the reasoning of such writers, why not think through the effects of power spatially?

It is this curiosity around all things spatial that I hope, in part II, will deliver a topology of power sensitive to the diverse geographies of proximity and reach. Rather than assume a landscape of fixed distances, well-defined proximities and effortless reach, in the second part of the book I hope to be able to show the difference that geography makes to the exercise of power. What we have lost in relation to power is not only the sense in which power puts us in place but also the mingled expression of succession and simultaneity, presence and absence, remoteness and proximity, that shapes our experience of power.

5

Power in its Various Guises (and Disguises)

Strictly speaking, Foucault's observations were broadly correct: power is nothing more than a series of effects, some of which (even if he did not quite express it this way) close down possibilities, manipulate choices, threaten violence, seek compliance or even work to crush, if that is not too literal a term, our free will. Yet for any account of power to remain plausible it has to take into account not only the manner in which it is exercised but also the basis of its production and the relationships through which it is mobilized. In that sense, it is appropriate to talk about arrangements of power as a term which allows for the possibility of considering *both* the generation of power through resource mobilization *and* the nature of its varied effects. For although much of what we understand of power stems from its relational effects, its basic configuration, whether mainly political, cultural or economic in shape, is drawn from the manner in which resources are mobilized and deployed over variable spans of space and time.

It is, none the less, the (inelegantly termed) *relational effects* that give their name to what most of us would have little difficulty in recognizing as power: namely, that brush with some awkward bureaucratic manager or some other notable figure of authority, or that feeling of deception which accompanies an act of manipulative advertising, or the erosion of consumer choice that emanates from the spread of corporate domination. Besides, it is only through the effects of such relations that it is possible to know and experience what it means to be on the receiving end of an act of power. Much of this, it seems to me, is not well understood, and in part we can trace this to the mix of insights and failings that characterize the broad accounts of power outlined in part I.

My aim in this chapter, however, is not to reject all that has gone before on the topic of power and space, but to *turn it to new purposes*; indeed to recast it in ways that I suspect many a time-honoured theorist would baulk at or would be unlikely to entertain. After a preliminary outline of how I understand the spatial constitution of power, I take the legacy of thinking about power as a circulatory medium that runs from Parsons through Giddens and Mann and includes, at a pinch, the recent writings of Castells, and recast it as an account of resource mobilization in a manner that perhaps even Mann would not altogether anticipate. In particular, I argue for a sharp distinction to be drawn between the exercise of power *and* the resources and capabilities mobilized to sustain that exercise. Following that, I try to live up to the legacy of Weber and Arendt by exploring some perennial concerns about power and its *modalities*, but in a context where their changing composure in space and time would not have embarrassed even Foucault. Nor, I hope, would the manner in which the distinctive modalities take effect in instrumental and collaborative arrangements of power.

First, though, I want to spell out what I take power to be – in its various guises and, for that matter, its disguises.

Opening Up Power...

Contrary to the impression given by 'centred' notions of power in chapter 2, namely that power is something capable of being marshalled and possessed, there are no pre-formed blocs of power out there waiting to constrain or limit the choices of others. There are only resources and abilities of many different kinds, ranging from varying amounts of money, property, land and goods to those that are less quantifiable, such as position, rank and influence of some sort, which *may* be mobilized and deployed to produce what we would recognize as power. Whether or not such mobilizations are effective is, of course, a different matter. For present purposes, we need only note that the distinct separation of something called power from the resources which are capable of generating it (as an effect) is one of a number of conclusions drawn by both Giddens and Mann, following the lead of Parsons, in their respective theorizations of power outlined in chapter 3. The idea that it is a misnomer to refer to power itself as a resource considerably extends our understanding of the nature of power, as indeed does the further point that resources are in fact only the media *through which* power is exercised.

Despite this advice, however, Giddens and Mann's analyses, along with those of theorists who adopt a similar mobilizing conception of power, are markedly one-sided in their treatment. What is gained through an understanding of the production of power and the specificities of resource mobilization is undermined by the lack of attention paid to the mediated exercise of power. Resources, for someone like Mann, may move through institutional networks in all kinds of interesting ways that combine authoritative and diffused techniques to achieve some far-reaching or proximate goal. Yet unless we are mindful of the mediated nature of power, wherein the effects of power may be modified, displaced or disrupted depending upon the relationships that come into play, the analysis falls some way short of any developed sense of how power may actually be exercised.

What do we mean by the *mediated* nature of power in this context? In many respects it is easier to say what the term does *not* refer to before we strike a positive reference.

In the first place, we should be wary of any mediated exercise of power that implies some kind of continuous transformation model of power as always in play. This seems to be what Bruno Latour (1986) has in mind when he speaks of a 'translation' model of power where everyone shapes the overall process according to their own interests and preferences as the 'order' is passed down the line, so to speak. On this view, the idea that power is transmitted intact from point to point is quickly rejected in favour of a transformational image that depicts a chain of agents composing power at every turn, as if it were an infinitely malleable substance. The sense in which the mediation of power must *always* involve alteration or change is not a particularly helpful one, however.

Nor, for that matter, is the view that mediation is essentially about an array of contingent circumstances under which predisposed powers may or may not be realized. If that were the case, we would be back in the landscape of chapter 2, with its pre-formed powers and latent capacities waiting to see whether or not the exercise of power by some dominant type was realized successfully south of the border or north of the border, east or west. The idea, however, that an exercise in mediation amounts to little more than an afterthought to a power that is always already there and waiting to be used is a peculiarly impoverished one.

Both these ways of thinking about power and mediation – it is always changing or it is always contingent – have run their course. They continue to shape and influence many of our expectations as to what power is: at one extreme, as a kind of plastic force capable of transforming itself at every turn, and contextual to the very end or, at the other extreme, as a rather solid capability waiting to be called upon to exert all manner of pressure. But if we remind ourselves of the need to keep apart the process of resource mobilization from the effects that amount to power, we can avoid both extremes. We can do this by thinking about power as just so many arrangements, each comprising any number of *resources* – ideas, expertise, knowledges, contacts, finance and so forth – which are mobilized to produce a succession of mediating *effects* in space and time which play across one another

The advantages of this formulation are, in my view, twofold.

First, it allows us to distinguish between resources and effects in terms of their spacing and timing, rather than consider them as all of one piece. In this way, the effects of power may be considered as spatially and temporally discreet from the resources that are mobilized to produce them. The imposition of a dominant or an authoritative presence may take place through a succession of episodes quite separate from those that involve the drawing together and settlement of resources. As such, there are no pre-formed powers acting from afar, just a blurred succession of mediating relations by which one finds oneself constrained or enabled, or indeed both.

Second, the formulation still makes it possible to consider power as coextensive with its field of study, be it the institutional reach of a government agency, the confines of a prison setting, or the scope of, say, a suggestive advertising campaign. In the case of the first, a professional grouping intent on enforcing a certain standard of welfare provision, for example, may involve the persuasive acts of top officials attempting to set down overall limits and expectations, the imposition of contractual constraints by regional managers, the delegated authority of supervisors and practitioners, and attempts by front-line agencies to reward clients for their compliance. All that this really means, in fact, is that there is a different interplay of forces to be found in various settings, at different times, in no particular order, which play across each other. Of course, the dispersal of responsibilities across an organization may lead to an independent use made of them, to the displacement of an original intent, or, quite simply, it may not. Similarly, in the case of an advertising compaign, the widespread diffusion of ideas which exert a clear pressure on others to conform may lead to their take-up in ways that are quite contrary to the initial objectives, or they may be perfectly in line with them. But that is the point.

There is no compelling reason why the mediated exercise of power should always involve its transformation at every twist and turn, since an acknowledgement of mediation is itself only a recognition that the exercise of power holds out the possibility for its displacement *or* its stabilization across a range of settings. None the less, to recognize that power is *mediated relationally* is an important counterweight to those who stress some 'centralized' capability which radiates out intact, or to those who exaggerate power's effectiveness by 'calculating' its impact from a given set of practices.

...And Narrowing it Down

As previously noted, the recognition that power is coextensive with its field of study was one of Foucault's strengths. While that particular insight does not amount to a striking departure from all previous conceptions of power, the idea that as free subjects we choose to govern ourselves by assuming the shape that is laid down for us does represent a break from the more familiar prohibitive models. The sense in which power functions through people 'working' on their own conduct, fashioning themselves in ways that reflect their acceptance of the 'truth' of an arrangement which, unwittingly, makes them subject to its control, neatly encapsulates the constitutive nature of power.

From this it is possible to imagine how various sets of techniques and practices could be brought to bear in the above example of welfare standards and their enforcement, from laying down guidelines as to the appropriate way of doing things (how to calculate, supervise, train, evaluate, grade, etc.) to the diffusion of ideas and received 'truths' which serve to justify privatized actions, and so on. Indeed, it is not hard to imagine how such a display of power could work in practically any setting, from the marketplace to the workplace, or from the schoolroom to the bedroom.

Yet, in fact, that ease of application is precisely the problem. For if one of the perceived strengths of Foucault's analytics of power is its breadth, it is also one its weaknesses.

The issue comes back to a point made earlier in respect of Foucault's treatment of power, and Deleuze's related appreciation of it: when any kind of convention, procedure, guideline, suggestion, influence or provocation is deemed to count as a technique or relation of power, there comes a moment when, rather disconcertingly, everything begins to shade into power. Moreover, the movement is one of surprising ease. In an otherwise lucid study of Foucault's work, Alan Sheridan, for example, describes power as:

> an effect of the operation of social relationships, between groups and between individuals. It is not unitary: it has no essence. There are as many forms of power as there are types of relationships. Every group and every individual exercises power and is subjected to it. (1980: 218)

Or here is Foucault himself, in his insightful essay on 'The subject and power', in which he argues against an overarching, 'centred' conception of power:

> This is not to say, however, that there is a primary and fundamental principle of power which dominates society down to the smallest detail; but, taking as point of departure the possibility of action upon the action of others (which is coextensive with every social relationship), multiple forms of individual disparity, of objectives, of the given application of power over ourselves or others, of, in varying degrees, partial or universal institutionalization, of more or less deliberate organization, one can define different forms of power. The forms and the specific situations of the government of men by one another in a given society are multiple. (1982: 224)

Foucault goes on to describe how such forms are superimposed one over the other and may cross or reinforce one another's effects, or at times even cancel each other out. To admit such a multiplicity of forms, however, to assert as Sheridan does that there are as many forms of power as there are types of relationship, is to ignore what is *specific* to power as a social relation: *it is to lose sight of the different modalities through which power is exercised*. It is to minimize the feeling for what power is when we brush up against it, when we experience its restricted choices or feel sure that certain possibilities have been closed down around us. Or perhaps worse, it is to gloss over the repressive side to power which may be experienced in a direct, instrumental manner. What is needed, therefore, is to pin down the exercise of power, to narrow or rather to focus the inquiry on what I take power to be.

Power as always of a particular kind

Notwithstanding Foucault's (1982) distinction between the direct nature of violent action and the indirectness of power as a social force, or Barry Hindess's (1996) attempt to read domination and government as two distinct modalities of the exercise of power,[1] there is little in Foucault's nominalistic treatment of the subject that approximates to the kind of analysis that both Max Weber and Hannah Arendt performed on power. Both were more than scrupulous in their attempts to pin down power relations in their various guises. Both paid particular attention to the different modalities through which power may be exercised and the realities to which they correspond. Which is not to say that the two of them delivered the last word on what is and what is not power, as indeed previous chapters have demonstrated. Rather, it is merely to state that Weber and Arendt, each in their own inimitable fashion, took particular care to examine and isolate the meaning and the use of the varied modes of power which underpinned the maze of practices in their diverse and wide-ranging studies.

---

1   Foucault (1982) contrasts violence with power to highlight the direct, immediate nature of violence and, like Arendt (1970), suggests that violence is not a relational act. Arendt is more forceful in this respect, asserting that violence is the demonstration of an ability which, although proven in relation to others, is essentially independent of them, in so far as the act and its target may be arbitrary. Arendt's opposing of power and violence was spelt out in chapter 3. Hindess's (1996) distinction between domination and government has rather more to do with the shift in Foucault's thinking towards the role of the self in regulating conduct than any specific attempt to pin down modal forms of power.

On a number of occasions in this book, I have drawn attention to the specific qualities that distinguish one kind of power from another; qualities which should not be conflated if we are to maintain an empirical grasp as to what power can do. Domination as a specific modality of power was introduced in chapter 2, primarily through the detailed work of Weber, who, it was pointed out, was at pains to distinguish principally between domination and authority, especially in his organizational studies. Care was also taken to demonstrate how seduction, as a particular kind of power, worked within a framework that is almost the mirror image of domination, where the possibility of indifference or outright refusal is integral to its very mode of operation. Likewise, coercion, as the threat of force or negative sanctions, was stressed in part I, as indeed was its more positive counterpart, inducement, through which people are won over to the advantages of something and bring themselves into line. Both sets of relations involve quite specific ways of exercising power which, I would maintain, entail only *certain* practices and techniques in *particular* modal arrangements.

Thus there is little to be gained by assuming that all practices reveal a kernel of power: that there are as many kinds of power as there are types of relationship. Of course, Sheridan's remark may have been of a more casual, off-the-cuff nature, which suggests that to take him literally is to pursue a false trail. Yet such casual understandings are not without consequence.

For nominalistic approaches of this ilk, at a time when, as we saw in chapter 4, the molecular and the multiple occupy the intellectual horizon, lend themselves to the exploration of singular instances of power. A concern for the singularity of individual techniques of power is perhaps another way of putting it. For a whole host of reasons, such a concern is perfectly reasonable. There are moments when it is entirely appropriate to focus on a 'bundle' of practices or techniques which may indirectly limit our behaviour in certain ways. Certainly (and quite correctly) there is little inclination among nominalists like Foucault and others to consider that bundle as adding up to anything more general. Yet in rejecting any general 'thisness' to power and sticking steadfastly to the singular they fall prey to a related misunderstanding, which is to think that there are no *specific* guises that power can possibly take.

It perhaps needs to be repeated that modalities of power are not entities that can be understood by breaking them down into an endless stream of practices which are then added together again to realize the whole. Whether it be domination or coercion, seduction or inducement, manipulation or authority under the microscope, each guise possesses its own empirical logic, its own specific qualities that mark it out as a particular modality of power. The realities to which they correspond should not be obscured by a worthy concern to delimit the multitude of practices and techniques which underpin the exercise of power, although unfortunately this is all too often the case. Arendt was indeed correct to insist that we should pin down the specificity of power relations and not reduce them simply to the 'business of domination' or extend them to encompass all types of social relationships (1970: 44).

Power as always already spatial

In the light of the foregoing, I would press the point that the modal qualities of power are not synonymous with the more commonplace practices which shape much of what goes on around us in daily life. Although the two things, modes and practices, are not unrelated, a concern for the former directs attention towards the substantive qualities of power which tend to be glossed over in more nominalistic accounts. In doing so, however, we need to take care that we do not throw away all the spatial vocabulary that comes with Foucaldian-type approaches, in particular their sensitivity to the spatial and temporal constitution of power. If the exercise of power, as I have argued, is always mediated in space and time, then so too are its modal qualities.

Something like seduction, for instance, does not begin and end its institutional life *as* seduction: to win over the hearts and minds of consumers in new markets, for example, involves the mobilization of various material and human resources by a range of economic agencies, the development of a strategy to raise awareness of specific brands and products, the diffusion of certain lifestyle ideals that are presented as within reach, the targeting of certain groups, and so forth. The effects of such a campaign, however, may be experienced by consumers as *manipulation* on the part of an untrustworthy corporation, or as an attempt by an outside agency to impose its voice and *authority* on the people, or as an act of *domination*, where choices are quickly narrowed down. Or it may resemble the modest form of power that bears the qualities of *seduction*. The point is that such final effects cannot be known and registered in advance, for at each twist and turn the exercise of power may (or may not) take on a different shape. The fact that the exercise of power is always and everywhere mediated implies that we have to take its spatial and temporal constitution seriously, regardless of the fact that it does not always make a difference in specific concrete circumstances.

Now, at this point, I wish to broaden the argument in a manner that is perhaps more controversial. In much of what follows and especially in subsequent chapters, I will argue that *spatiality is constitutive of power relations not only in general, but also in the particular ways in which different modes of power take effect*. This involves thinking through aspects of spatiality beyond that of a confined setting or context to consider the diverse geographies of proximity and reach, and in particular what it means to exercise power 'at a distance'. Some, but by no means all, of these issues have been touched upon already. Much of what I wish to say, however, has less to do with the idea of spatial power plays – a little stretching of the resource base here, an extension of political and economic influence there – and rather more to do with the effective reach, as well as arrangement, of different modes of power.

Earlier, for instance, we spoke about seduction as a modest form of power which, following Gilles Lipovetsky (1994), has that characteristic precisely because it acts upon the choices of those who have the possibility to opt out. What Lipovetsky did not go on to consider, however, in part because it is not easily brought within his narrowly historical embrace, was that the very indeterminacy he identified as an attribute of seduction is also what enables its exercise to be effective 'at a distance', in terms of its reach. And it is effective in this way by virtue of its limited effects – suggestion rather than prescription – which are an integral not a residual feature of how it is exercised. The fact that seduction works on curiosity, seeking to take advantage of attitudes and values that are already present, leaving open the possibility of rejection or indifference, is what gives it its considerable reach, yet at the same time curbs its intensity. These are not accidental features of the way in which seduction works; they are qualities that distinguish seduction from other modes of power and mark it out as a distinct way of exercising power.

In due course, we will look at a range of different modalities of power, some instrumental in their leverage, others more collaborative by design, with an eye to the varied nature of their spatial qualities. However, once we begin to rearrange power in this way, as a series of modal effects for which space makes a difference, we need to be alert to the pitfalls of spatial determinism: the idea that space itself is what lies behind the exercise of power regardless of the distinctive attributes of seduction, domination, authority, coercion or whatever. This is the controversial side to the claim that spatiality matters.[2]

In order to avoid misunderstanding, we need, first, to be clear about the fact that when something like a global advertising agency works on the choices of consumers what actually happens *depends* upon the relationships

---

2 The pitfalls of spatial determinism stem in this instance from the danger of attributing causal characteristics to spatial qualities regardless of how something like seduction as a mode of power could possibility have an effect. While it is feasible to abstract the spatial form of seduction as effective 'at a distance', little or nothing can be said about how it makes a difference in terms of reach. Distance itself, the simple movement across space, provides no clue as to anything unless the power relations involved and their characteristics are considered. For an illuminating discussion from a realist perspective on what space can and cannot do, see Sayer (2000: 108–30).

they encounter and the resources that others bring to bear. A modest form of power like seduction, under certain circumstances, may take on an uncharacteristic depth, especially if the audience largely comprises what may be referred to rather disparagingly as cultural dupes. Or its message may simply be rejected or ignored by those not won over by the nature of the superficial appeal, perhaps because they have the knowledge and information resources to think otherwise. In either case, seduction remains a rather indeterminate mode of power with considerable reach and limited intensity. For that is merely the result of seduction being the kind of power relation that it is and not in the same mould as domination or manipulation or authority. While we cannot and should not insist that seduction will be effective across a range of circumstances or that it will undoubtedly work at arm's length, we can none the less say something about the way spatiality is constitutive of seduction as a distinctive exercise of power.

Perhaps the key point to note in all this is that there is *no* spatial template for power. The dispersion of power over more or less long distances may actually problematize the establishment of a powerful presence in key locations, for instance. Yet the speed-up of communications, the rapid circulation of people and information, and whatever else is required to mediate power relations across space may overcome any such problems and actually enhance the control and reach of institutions and agencies. Distance cuts both ways. But even to acknowledge this does not mean that we can afford to gloss over the spatially circumscribed nature of power relations. For unless we consider the kind of power we are talking about, the relationally and spatially distinct modes in question, we fail to grasp the *particular* ways in which modes of power take effect.

In one sense, what is involved is the development of a distinct spatial curiosity: an attempt to rethink spatially the modalities of power deployed in many a traditional account, especially those presented by such figures as Max Weber and Hannah Arendt, and to go beyond them in terms of our present spatial vocabularies of power. As I see it, while the writers discussed in part I often represent spatiality in rather a stilted fashion, they do none the less share a sense that spatiality has some significance to the workings of power. For some, that significance is characteristically vague, whereas for others, it is all to the fore in the nuances and the detail. It might be more rewarding, however, *to think through the effects of power spatially rather than to concern ourselves simply with documenting the spatial aspects of power.* Perhaps it is time to remind ourselves of the fact that power relations have long been experienced through a diversity of different modes and that they are always already spatial.

We can start this process by, first, understanding *why* it is that we frequently mistake resources for power.

Power in Name Only, or Mistaking Resources for Power

Resources, to reiterate a point made earlier, are the media through which power is exercised. To talk of power as a resource, then, is a misnomer. And yet most of us do it all the time, with, it would seem, good cause. For there is something about the language of resources and the way in which they are used that lends itself to the idea that resources *are* power. There is after all, something deeply familiar about the view that money is power. A rich person is powerful, everyone knows that and that is all there is to it. Thus when we speak of, say, knowledge as an instrument of power or wealth as the basis of power, it is relatively easy to find oneself in a vocabulary that slides effortlessly between resources, instruments, capabilities and power. Although the terms do not quite work as synonyms, they do translate from one to the other with relative ease. A possible reason for this is that the exercise of power, it seems to me, is often observed 'after the event', so to speak. When we look back upon a particularly crushing act of domination, for example, by, say, a giant 'rogue' corporation or a particularly wealthy individual, it is relatively easy to read it as the almost inevitable outcome of their 'size', as measured by the vast resources at their disposal. Thus we conflate power with total resources; we view the outcome as the result of the institution's or the individual's undivided capacity and fall back into the kind of quantitative capacity position criticized by Hindess (1996).

Yet there is no real justification for this slide between power and resources. From the standpoint of the here and now, observing events as they unfold, a range of outcomes may appear possible, including the misapplication of resources by well-intentioned but basically misguided actors or, worse, the gross miscalculation of resources available (either under- or overestimation; it makes little difference to the point). In some cases, incompetence or ignorance may even have to be factored into the equation. When you are caught up in the process like this, there is no telling exactly what form power will take and how effective its execution will be. In other words, the exercise of power and its mode of operation cannot be 'read off' from the resources available, whatever their size or magnitude.

The two things, then, the exercise of power and the resources mobilized to sustain it, are not of the same order. So why has such a misunderstanding arisen? There would appear to be at least three possible reasons to consider.

First, there is, as indicated, the tendency to adopt a retrospective view of the exercise of power and to attribute the outcome to what seems like 'power in things', which are really nothing more than the resources and abilities mobilized. This attribution is aided and abetted by the fact that in the English-speaking world we use the *same term*, power, to refer both to the capacity to do something *and* to the act itself – of domination, of authority, of whatever – which may flow from the resources mobilized. The French language, by contrast, has two terms to decipher power, *puissance* and *pouvoir* (as indeed does Italian, *potenza* and *potere*, from the Latin base, *potentia* and *potestas*), where broadly the former denotes capacity and the latter denotes the act of power. This helps to minimize possible confusion.

Second, there is the inclination, following on from the above, to treat 'power in things' as solid, almost permanent by nature. The notion that power may be held 'in reserve' by institutions like the big media or software corporations or is something that people 'cling on' to is an extension of this way of thinking, and, far from being a caricature, it has the potential to obscure the fact that both resources and power are reproduced over space and time. For there are no powers that 'belong' to anyone on an indefinite basis.

And third, the conflation of resource size with power can be traced in part to a fairly common spatial misconception. To speak of people or institutions 'having' power, for instance, is to return to the vocabulary of chapter 2 and the idea that power is centralized in such bodies. Bill Gates 'holds' power and it is centralized in Microsoft corporation, or so it would seem. But Seattle and Microsoft no more 'have' power than Bill Gates 'holds' it; yet they do possess a rich concentration and mix of resources and abilities on the basis of which power is exercised.

In short, there is no need to endorse a scenario of 'powerful' bodies, each with its own capacity for domination held 'in reserve'. Such a picture is misleading. Yet we cannot simply wish it away, as if power were not commonly thought of as something which is struggled over, won and lost. We need to delve a little further into the reasons given above and to probe their hold on our imagination before we move on.

Prising power and resources apart

The bottom line is that when the term 'power' is used to describe the resources mobilized by people and institutions it is being applied to something which, without my wishing to legislate on language, it does not denote. Resources are part of any arrangement of power, but they do not denote its exercise.

Given the slippage in vocabulary between resources, instruments, capabilities and power that often seems to occur, it is perhaps useful at this point to recall Arendt's firmly held view that power should never be seen as the property of any one person or thing; rather it should be recognized as something which arises through mutual action, through the collective mobilization of resources which provides the basis upon which action to meet agreed ends takes place. On this understanding, failure to act together or the inability to pool resources effectively results in the evaporation of power. It simply fades from view or, alternatively, never sees the light of day in the first place. Whilst we need not fully endorse such an account, especially given its more solidaristic overtones, it does none the less draw attention to the often tenuous nature of resource mobilization and provide a clue as to its separation from the act of power. Power is exercised *in relation* to the resources mobilized and the two forms of social interaction need to be prised apart, both analytically and substantively.

As mentioned earlier, in French it is more or less possible to discriminate linguistically between resource capacity and the exercise of power, whereas in English (or German, for that matter) we tend to subsume both the general capacity and the particular act under the single term 'power' (likewise in German, under the term *Macht*). Raymond Aron (1986) has attempted to disentangle some of the issues by setting out the distinctions involved:

> French has two words to translate '*Macht*' or '*power*': '*pouvoir*' and '*puissance*'. Both have the same origin: the Latin verb 'posse' ('to be capable of, to have the strength to'). '*Pouvoir*' is the infinitive of the verb and according to Littré's definition, 'merely denotes the action', whilst '*puissance*' (the participle) designates 'something lasting and permanent'. One has the *puissance* to do something, and one exercises the *pouvoir* to do it. It is for this reason that we can talk of the *puissance* of a machine, and not its *pouvoir*. (1986: 255–6)

Although far from watertight, the distinction in French does make it possible to think about, on the one hand, the resource bases or instruments which generate a capability for action and, on the other hand, the actions themselves. Both allocative and authoritative resources referred to by Giddens, for instance, would fall neatly under *puissance*, so that access to property, finance, knowledge and position would in combination produce a general capability for action. Yet *puissance* also seems to allow for more than just that. Power is still wrapped up with the meaning of *puissance*, largely it would appear through the inclusion of some kind of potential inherent within an object's or an institution's capabilities. As Aron goes on to outline in respect of the *puissance/pouvoir* distinction, it:

> would therefore be roughly that between potential and act. It is not unhelpful to recall that in English 'power', depending upon the particular case, is applied to either a 'potential' or an 'act'. Or it designates a potential that is revealed and whose scope is measured by passing to the act itself. (1986: 256)

On this interpretation, therefore, *puissance* amounts to more than a bundle of resources which, when drawn upon, may lead to an act of power; it also refers to a potential capability. The language is again reminiscent of chapter 2 and the view that certain things are disposed to act in certain ways by virtue of their constitution. This is fine, as long as we keep in mind that power is *not* in things and that any ability or aptitude is the outcome of an ongoing process of resource mobilization and not a substitute for it. A body may appear to 'have' the potential to do something – like a corporate body or a 'powerful' lobby group with extensive bargaining capacity in the political arena – but it is the assemblage of resources and abilities, *not power* as such, which gives rise to it. Once again, it is because power is frequently observed retrospectively, 'after the event', that the exercise of power is thought to be a manifestation of something deeper. At this point, we leave the messy territory of resources and step over into metaphysics.

In fact, this was precisely the kind of step that Foucault was keen to avoid. It is no accident that there is little, if anything, in his writings on power that resemble a theory of *puissance*, yet, as was mentioned earlier, he devotes considerable attention to the development of an analytics of power which conveys the ins and outs of the exercise of power (*pouvoir*). In his essay 'The subject and power', he takes the opportunity to set forth explicitly what distinguishes his account of power from those that have the apparatus of *puissance* as their starting point. Asking questions about the 'how' of power rather than attempting to seek a deeper presence, he goes on to elaborate the focus of his critical examination:

> 'How', not in the sense of 'How does it manifest itself?' but 'By what means is it exercised?' and 'What happens when individuals exert (as they say) power over others?'
> As far as this power is concerned, it is first necessary to distinguish that which is exerted over things and gives the ability to modify, use, consume, or destroy them – a power which stems from aptitudes directly inherent in the body or relayed by external instruments. Let us say that here it is a question of 'capacity'. On the other hand, what characterizes the power we are analysing is that it brings into play relations between individuals (or between groups). For let us not deceive ourselves; if we speak of the structures or the mechanisms of power, it is only insofar as we suppose that certain persons exercise power over others. (1986: 217)

I leave it to others to debate the textual ambiguities that such passages may contain. What is plainly uppermost in Foucault's mind, however, is that power amounts to a relationship which involves those on the 'receiving end' no less than those who exercise it. In the terminology of chapter 4, power relations are an immanent force in so far as power is inseparable from its effects, that is, from what it can actually do. There is no sense in which power remains latent, hidden from view, waiting to show itself when the time and place are right to make its presence felt. Thus there is no deeper potential out there marking time, ready to impose itself from the outside; for Foucault there are only power relations which are exercised on and through subjects. Capabilities, however, are of a different order; within any particular arrangement, they serve as the means to power's ends and, I would argue, provide constancy of shape.

This sense, therefore, is consistent with the view that power is something which is exercised in relation to the resources and abilities mobilized. It also has a strong emphasis upon action, upon practice, that takes us closer to the meaning of *pouvoir*; which in its various conjugations expresses a kind of 'can-do' mentality, an energy which is all about something coming into play in a positive sense.[3] From this, it is easy to see how an emphasis upon everyday techniques and practices by Foucault takes him closer to what he understands as power: namely, the ability of people to govern themselves by assuming the shapes indirectly laid down for them.

It may be helpful to round off this discussion of the *puissance* (capacity)/ *pouvoir* (exercise) distinction with an illustration of what is at stake if we do not hold apart resources and capabilities from power and how it is exercised.

Take the familiar example of Hollywood domination in relation to film and cinema. We touched upon this in chapter 2 with respect to the claim that Hollywood acts as a direct threat to the French way of life through the former's domination of visual culture. Now, if we roll together Hollywood's resource capacity and what its major operators actually do to get a picture overseas, we might easily regard Hollywood as 'having' the power to dominate French audiences by virtue of its financial muscle, production studio capabilities and extensive distribution networks across the globe. We could blur the arrangement further by talking about the sheer size of the major players in Hollywood, the concentration of control in the hands of a few multimedia transnationals, and the diversification of their interests across a range of different products. From this it would be but a short move to conclude that the extensive reach and high final market concentration among the majors can be explained by their dominant position within the global media hierarchy (as indeed Asu Aksoy and Kevin Robins 1992 do in their argument for Hollywood's ongoing domination; see also David Morley and Kevin Robins 1995, and Franco Moretti 2001). To do so, however,

---

3   Gayatri Chakravorty Spivak (1996: 141–74) captures this ordinary-language sense when she talks about putting some of the homely verbiness of *savoir* and *savoir faire* into *pouvoir faire* so that power, the act of doing something only as you are able to make sense of it, comes closer to what Foucault was trying to convey. My only reservation is that this way of thinking often assumes that inducement is the *only* mode of power possible in this analytic. This is a long way, however, from the apparatus of *puissance*, which has the feel of a power standing over others, as in the shape of the overarching infrastructure and capability of a nation state.

would be to read the situation backwards, from the acts of domination back to the size and capabilities of the majors – back to what so often passes for 'economic power'.

Alternatively, we could cut into the process at the points where power *is* exercised over others, along the length and breadth of the institutional arrangements of the major operators, to the ways in which choices are shaped in the final marketplace. This may reveal a more open, mediated process, from the simple fact that sections of the French public, despite the government's cultural policy, may have actively sought American cultural images and icons because these have been presented to them as an attractive option, to an altogether different context where French cinema exhibitors have colluded with distributors to show an unbalanced diet of Hollywood film fare. Neither case is one of domination or financial 'strength', nor do they amount to an exercise in cultural coercion where French audiences in Paris and Lyon are forced to watch a diet of US movies. If anything, the former case approximates to an act of seduction and the latter to one of manipulation. Yet neither of these effects would be apparent were it to be assumed that the exercise of power is merely a reflection of the size and capacity of the dominant players involved.

The important issue, then, surely is to consider the modes through which power is exercised separately from the process of resource mobilization, and not to fall into relaxed assumptions that resources and abilities equal power.

Reserves of power or settled resources?

A similar casual approach to resources and power is evident in the way that we habitually talk of 'powerful' institutions or of despots 'using and abusing' power *as if* they possess some kind of 'reserve' or 'store' of power which can be dusted down and drawn upon at will. There is a variety of odd yet familiar expressions which, in effect, serve to consolidate the impression that power may be 'hoarded' or 'accumulated' over time, such as the notion that power is 'lodged' or 'vested' in, say, the big film studios or that Los Angeles and Hollywood act as some kind of 'container' of cultural power. That Los Angeles and its studios possess none of these characteristics is almost beside the point when set alongside the creative assets, skills, knowledge and other resources that Hollywood so obviously has at its disposal. Power, as we know, is not the same thing as leverage over resources, yet it seems somehow churlish to criticize what may after all be little more than a series of overstretched metaphors. To speak of 'reserves' of power, however, is to do more than commit a simple misnomer.

Perhaps of all the metaphors that work to elide resources and power, the notion of 'stored-up' power is one of the few that has the potential to disarm those who wish to change the status quo. Although it has to be said that the notion hardly paralyses opposition, when confronted, for instance, by the apparent accumulation of economic and symbolic power 'deposited' over time in Los Angeles, it is not altogether surprising to find that challenges to Hollywood's reach are often characterized by a degree of fatalism. That events will take an inevitable course regardless of any intervention or protest is not an unexpected response, for example, from those in France who oppose the American dominance of the global media.

Much the same kind of resignation can be found in other areas where the legacies of 'power' are evident for all to see. The long-standing inequalities which structure relations between the many and varied global cities, Los Angeles and Paris included, are a case in point. For example, the so-called 'power-filled' spaces of New York's financial district or those of the City of London display (through their architecture and monumentality) the cumulative advantages of having been dominant for so long in the world of financial flows that there seems little point in challenging them or questioning the durability of their economic power. Similarly, for their less affluent counterparts in cities which aspire to be something more on the global economic stage, such as Malaysia's Kuala Lumpur or Taiwan's Taipei, the absence of any such legacy appears to be a permanent disadvantage. Power, in this fatalistic assessment, appears to 'belong' to certain cities and not to others, and makes it that much harder to contemplate an alternative global ordering.

What would Arendt have made of all this? For one thing, she would have rejected out of hand attitudes which express such passive fatalism. The idea that power 'belongs' to anyone or anything on an indefinite basis is simply not part of her understanding of power. For Arendt, there is a sense in which power is always of the moment, something that has to be continuously reproduced over time, especially if it is to be sustained in the form of a 'legacy'. The impression that power can be 'stored' within the boundaries of a city or wherever is a false one, gained in part by the fact that global cities like New York and London maintain their resource base through successive mobilizations. An existing position of economic strength and influence helps such cities, of course, and it would be foolish to deny it, but the manner in which they stabilize assets and resources represents an ongoing process of negotiation, enrolment and enactment which belies any sense that power is a permanent attribute of these locations.

Arendt was right, in my view, to stress the impermanent nature of power and the need to mobilize resources on a continuous basis. Where I part company with her, however, is over the fleeting nature of power asserted and the almost ephemeral process of resource production outlined. Both aspects are somewhat exaggerated in her analysis (largely one suspects because of the emphasis upon association and collaboration). It is one thing to point to the tenuous nature of resource mobilization and quite another to suggest that the whole arrangement is so fragile and fleeting that nothing is settled or fixed to any real degree.

After all, if we continue with the example of global cities, the attempt to hold down streams of assets and resources and to put them to work in constructive ways is precisely what financial analysts and investment bankers, lawyers and accountants, management consultants and others actually *do* in places as far apart as New York and London, or Frankfurt and Tokyo. Fixing resources in these commercial locations, to paraphrase Saskia Sassen (1991, 1994, 1995), is the task performed by such professionals in order to produce a capability for global control in the world's capital and commodity markets. What we can say with some degree of confidence is that the amalgam of critical service activities to be found in the central locations of such cities does not amount to anything like an everlasting capacity. But if there is no ever-present 'reserve' waiting to be tapped, equally the whole production complex in somewhere like the City of London is not just a fleeting arrangement put together for as long as the economic moment lasts.

What we need is a better grasp of how resources *flow* and how they are *fixed* in such locations, or indeed any location. For that, we can return to the work of Manuel Castells, which was discussed at the end of chapter 3. What Castells's account lacks through its impoverished conceptualization of power is partly made up for by his detailed consideration of networked flows and co-present interaction.

One of the key points that Castells wishes to argue – indeed, he often repeats it like a mantra – is that 'the power of flows takes precedence over the flows of power' in what he takes to be the onset of a 'network society' (1996, 2000). By this, I take him to mean that power is mobilized through networks; that it is what flows *through* networks rather than any embedded capability which empowers particular groups and the institutions of which they are a part. On this view, the movement and flow of information, ideas, symbols, money, technology, and the forms of co-present interaction (mediated through communication technologies) which lead to their exchange and assimilation are precisely what give the professionals mentioned above their privileged position in the network of global cities. If we resist the temptation to imagine that it is actually power which travels through the networks and concentrate instead on the fact that all kinds of resources may be mobilized along their span, then the bankers, accountants, lawyers and the like located in New York or wherever have the allotted task of settling and co-ordinating, orchestrating and redirecting the flows of resources around the networks. Global cities, on this understanding, act less as a magnet to attract a greater concentration of resources than as a throughput to enable resources to flow and settle where directed.

This is a very brief summary of Castells's flawed 'power of flows' notion, yet it does convey the sense in which mobile resources like information, knowledge, ideas, contacts and money may be organized and settled (although Castells appears to have less to say about immobile resources such as land and property). Not all bodies generate resources in this manner, but the ability to trap and to stabilize all kinds of resources on an ongoing basis gives something of the flavour of what is involved in the process of resource base mobilization and its reproduction.

Thus in place of a permanent store of power and its more deterministic overtones, it is possible to conceive of resource mobilization and reproduction as a fluid capability, as a series of negotiations and enrolments which have as their aim the holding down or settlement of resources in a particular arrangement. In fact, there is much here that chimes with Latour's (1986) relational account of networked associations, despite his rather unhelpful image of continuous network transformation noted earlier. Indeed, Latour (1991), with Michel Callon (Callon and Latour 1981; Callon 1986), has done much to show how both the fixity and fluidity of relations represent a key starting point for thinking about resource mobilization and relationally constituted power, as we shall observe in the following chapter. For present purposes, the above understanding is enough of an antidote to notions of 'stored up' power or powers 'held in reserve' for us to move on to consider a further way in which resources are often mistaken for power.

Centralized power or the territorialization of resources?

Perhaps it is because we are so used to thinking about where power 'lies' – that the constraints upon us must after all emanate from somewhere – that it is easy to conceive of power as a centralized force from which all manner of rules, regulations and constraints ultimately stem. If it is not the charismatic authority figure so prominent in western culture or the overbearing executive situated at the apex of a corporate hierarchy, it is the monumental architecture of the old and established bank buildings, town halls and government ministries, or their more modern counterpart, the towering skyscrapers that dominate modern city landscapes, which serve to reinforce the impression that power can be identified and traced back to its point of origin. Such icons and representations are the very stuff of centralized power; for what may well be a rather modest, insubstantial force, they provide the necessary materialization which speaks volumes about where 'power' ultimately appears to rest. And in that they are not wholly wrong.

Instead of dismissing such representational figures and images out of hand as if they crudely mask a real dispersion of power, we need to address the work that such representations perform and the basis of their enduring appeal. The interesting thing about such representations is that they do amount to an exercise of power and yet, at the same time, point to its material basis. An illustration from the work of Richard Sennett (1990) provides us with a sense as to how this is achieved and what is involved.

In his book *The Conscience of The Eye*, Sennett draws attention to the monumental nature of many of the buildings which comprise the New York skyline, and in particular to the towering form of the Rockefeller centre, built in the 1930s and 1940s. It would be simple to conclude that the centre represents a space of power, a mark of domination amidst a welter of modern symbolism, and leave it at that. But Sennett wants to argue that the Rockefeller centre represents a space of authority, not domination; the sombre nature of the buildings, the precise ordered spaces, and the sense of detachment from the humdrum of daily life around them symbolically stand for a guiding present, a source of protection and trust in an otherwise unsafe world.[4] This, then is not just another description of a so-called 'power-filled' space, but a dissection of an *act* of power. The image presented through the clear lines of the buildings, the materials chosen for their construction, the play of light and depth, amount to an exercise of power on the part of the architects and planners, which has now passed (non-human-like) on to the buildings themselves.

But, moving on from Sennett, that is not all that buildings of this monumental type represent. Sites like these, together with public and private buildings such as law courts and corporate headquarters, also speak to a certain territorialization of resources and organization. They point both symbolically and materially to the allocative and authoritative resources which underpin the power exercised by such organizations and the people who occupy positions of authority and control within them. Giddens, you may recall from chapter 3, was saying something similar, although for different reasons, when he pointed out that for resources to work on a systemic level they must be taken up and reproduced through

---

4   The sense of authority Sennett finds here owes much, it would appear, to Arendt's classical account, where authority is interpreted more as guidance and less as command. This interpretation is developed later in this chapter and was signalled earlier in chapter 3. In this instance, it is as if the weight of the past, the respect that comes with time, is accorded to the buildings themselves, the layout and design. Sennett (1980) considers at length paternalistic and instrumental styles of authority, with an eye to understanding the felt nature of the personal bond that arises through such relationships.

social practices on a routine basis. The routine fixing of resources, the ability to concentrate expertise and institutional authority and maintain them on a regular basis, does not take place 'on the wing', so to speak, in a world of constant flows; it takes place at particular sites which, as implied by Sassen, are *more or less* embedded within organizational complexes, hierarchical arrangements and institutional centres.

The stress that I am placing here upon the 'more or less' is quite deliberate. At the risk of belabouring the point, the sites in question are *not* centred locations from which power is amassed and then redistributed. There are no organizational outposts of the likes of Bill Gates's software empire (or the Rockefeller financial estate for that matter) standing by to receive discrete bundles of power and domination from an omnipotent centre. There are, however, situated and embedded arrangements which, for want of a better way of expressing it, represent clusters or sites at which the settlement of often fluid capabilities is a necessary pre-condition for the exercise of power. Some resources, such as the infrastructural resources of the state described by Mann (1984), namely taxation, law, property and the like, are more grounded, whereas others, such as finance, information, ideas, people and contacts, appear less fixed by nature.

It is, however, probably a mistake to generalize about the mobility of resources in this manner, as different sites – from Sennett's spaces of authority and the corporate headquarters of the big media companies to the public sites of collective action and distributive struggle – involve a tangled arrangement and concentration of positions, abilities and infrastructures. None the less, resources that combine, grow and change in unanticipated ways according to how and where they are settled rarely, if ever, exert complete and utter control from one site (or centre). Mobilization means just that: putting into readiness a capability for action, and in that sense many a territorial site is caught up in the fixed and fluid nature of power arrangements, much of which, as Deleuze and Guattari in *A Thousand Plateaus* style might say, eludes it, some of which transforms it, and without which the whole ensemble would not hold together.

This is not, however, to side with Castells (2000), who goes so far as to suggest that the new networked morphologies 'dissolve' centres, with the 'power of flows' (the information, knowledge, ideas and the like circulating in networks) bypassing established political centres and materialized sites of power. The state (and, one supposes, many other embedded institutions) is transformed in this scenario from a centralized power to a power-sharing node whose function is to switch, direct and co-ordinate the flows of exchange and interaction. There seems little point in engaging with such an exaggerated claim, but it may prove useful to endorse the problematization of centralized power that it raises, although not for the reasons that Castells has in mind.

To be clear, I wish to argue that the term 'centralized power' is yet another misnomer. Power, unlike resources, is not something that is fixed centrally, just as it can never be 'stored' like some fixed deposit. Of course, it often *seems* as though power is located in some central figure or institution, but that is principally because it is represented as such. What such representations actually point to is a resource base that has been stabilized over time; a capability or series of capabilities which have been produced through an ongoing process of mobilization. Resources, in short, are territorially embedded, but they are also mobilized through networked relationships. The two aspects, the fixture and fluidity of resources, may appear to be in contradiction, but it is the very tension between them that injects the dynamism into what would otherwise be a rather solid, staid power arrangement.

Resources, however, are only one part of an arrangement, the full meaning of which is exhausted only through the act of power itself, whenever and wherever it is exercised.

Recognizing Power, or Living Up to Weber and Arendt

So when we say that certain prominent figures, like heads of corporations or well-placed state officials, 'are powerful' or 'hold a powerful position', what we actually mean is that they can muster enough resources to have an *effect*. It is the effect that matters and, at the very least, is what we experience when power is exercised over or with others. It is this experience that is sometimes glossed over when 'powerful acts' are explained retrospectively, as if they emanate from some deeper, more intractable presence. In contrast, from the vantage point of becoming, when power acts so to speak, it is possible to get closer to what it is exactly that is being exercised. The stress upon *what* is exercised rather than upon *how* it is exercised may appear rather odd at first glance, but it has its advantages.

Principally, it directs attention to the kinds of interaction or modes of interplay involved, rather than to the means by which power is exercised, namely the techniques and practices which shape all manner of conduct. This is part of the process of narrowing down in focus referred to earlier, where the specific qualities which mark out sets of power relations are acknowledged and their meaning carefully assessed. In other words, it is to recognize power for what it is, not simply for how and by what means it is practised.

A case can be made, for instance, that not all rules, conventions, strategies and tactics are reducible to relations of power. More pointedly, as argued earlier in the chapter, against a Foucaldian-inspired approach a case can be mounted which suggests that focusing upon endless practices which appear to be combined in just about any permutation is the wrong analytical path to go down. For sure, such a path can take us forward and reveal the detail of the route, but it also rather confusingly takes us back to the starting point and the question of where power begins and where it ends. Besides which, the journey itself makes it that much harder to distinguish effects from practices, and to disentangle instrumental and collaborative modes of power from the multitude of techniques which fall under the route heading of government.

Modalities of power such as those that we have had cause to discuss, like domination and authority, seduction and manipulation, among others, are not easily understood by assuming that they more or less map onto the practices which guide and shape the conduct of daily life, whether at home or work, at a place of entertainment or leisure, or incarcerated in some confined institution. There is no one-to-one relationship between domination and a series of rules or conventions in the marketplace, for example, where it may seem to be such a simple matter to attach the label 'domination' to a case of restrictive practices or one of restricted consumer choice. Equally, we cannot simply add up a series of threatening practices and call it coercion, or take apart the representational practices of a pervasive discourse and pin the label 'inducement' on it. Such instances are evidence, in my view, of loose thinking and easy abstraction, but more importantly they are at fault for giving a name to something which holds a different meaning.

Modalities of power are not of the same order as a series of individualizing practices, regardless of whether normalization is their ultimate aim or not. There is no whole–part relationship involved where, say, authority is deemed to represent the sum total of apparent pieces of expertise or a range of compliant behaviours. Authority as a *specific* relation of power cannot be summarized in this way, or adequately understood for what it is: that is, as a modal effect of power with distinctive characteristics and circumscribed consequences. This is not to say that authority is always of a certain style, as any attempt to live up to Weber or Arendt would warn us. It is, however, to suggest that we should focus on *what* is exercised, on the modal effects of power, and not be distracted by the multitude of means through which power is exercised. Only then will we be in a position to grasp the distinctiveness of power and its modal arrangements.

Instrumental power, or power over others

Weber appreciated, perhaps more than many since, that where resources are unequally distributed between groups, the pattern of conflict likely to emerge will involve one side subjecting another to its instrumental will. The competitive and conflictual nature of power arises from this understanding, with power often reduced to a game of winners and losers as one side gains at the expense of the other. This zero-sum sketch of power was outlined in chapter 2 and certain reservations stated about the fact that the objectives and interests of different bodies should always be regarded as mutually exclusive. Rather than considering power as a potential force for integration or enablement, as discussed in chapter 3, or suspending judgement until it is apparent who is doing what to whom for what reasons and in which context, this instrumental view sees power as simply about bending the will of others. It is about exercising power over others. And of course Weber was right in many respects, certainly with regard to his studies on organizational structure and power.

The prime example, which we will not dwell upon in any depth, is obviously that of modern bureaucracy, with its well-defined hierarchical decision-making processes based upon impersonal rules and specialist knowledges.[5] Neither dominating in the general sense, nor coercive in the last resort, bureaucratic organizations for Weber are essentially maintained through relationships of *authority*. People, public and private officials alike, are placed in positions of authority, and by virtue of their institutional role others willingly accept their right to issue directives and to enforce an organization's rulings. Or so we are led to believe. Clearly, the willingness of others to comply with the demands of those in authority may not always be as forthcoming as it might appear to those looking in from the outside. Compliance is always conditional, and this is deemed to be met by organizations' clothing their actions and judgements in an anonymous set of rules and procedures. It is the impersonal nature of the position, the ability to divorce the role from personal ties and influences, which provides those in authority with the necessary legitimacy to exercise decision-making powers.

In this rule-bound environment, however, we need to be sure not to confuse the *position* of authority with the *exercise* of that authority, as the two are of a different order. Whereas the former is an institutional resource fixed within an organizational structure, the exercise of power by people *in*

---

5 Modern bureaucracy is, of course, a specific organizational structure from which Weber was careful not to generalize. Power is exercised over others in bureaucratic structures through the asymmetric relationships inscribed in a vertical chain of command. See Weber (1978: 217–26, 956–1005) on the administrative logic of modern bureaucracy and the pure type of rational-legal authority associated with its instrumental operation. As his is a centralized view of power, however, it is easy to overstate the effectiveness of this machine-like application. Barry Barnes (1988), among others, has pointed to the less than perfect authority in such command structures, where delegation inevitably leads to the use of discretion and the possibility that an independent use may be made of such 'powers'. This more provisional sense of power is developed further in chapter 6.

authority involves a series of relationships between those in the vertical chain of discipline and control. Officials may hold a bureaucratic position and delegate responsibility for decision-making downwards through the ranks, but the outcome of any act of authority is likely to be far more ambiguous than anything inscribed in the rule book.[6] A command–obedience type of relationship may be customary in military settings, but the ability to secure assent is by no means as straightforward, say, among administrators in a hospital or a private nursing home or their professional equivalent in a software or media company.

The point of all this is to register what *is* exercised. In many organizations like those mentioned, authority, as was pointed out in chapter 2, may only be seen to be exercised if it is recognized without question. Recognition holds the key to authority as a mode of power and others comply because of the acknowledged expertise or competence, for instance, of those who actually 'hold' authority. In a real sense, the act and the person are inextricably woven together when authority is exercised (in much the same way as the buildings and the act were woven together in Sennett's spaces of authority). There is, as Thomas Osborne (1998) has pointed out, an ethical dimension to authority whereby those who hold it must first bring themselves to order before they can bring others into line.

The integration of act and person in the exercise of power is even more apparent in the two remaining styles of authority explored by Weber. If modern bureaucratic rule-bound structures are associated with a rational, legal style of authority, then what counts as authoritative in patriarchal and patrimonial structures, or in exceptional moments of economic and political transition, is traditional authority and charismatic authority respectively. Both styles of authority are more overtly personal than that exercised in bureaucratic administrations, the former because it is rooted in traditional bonds of obedience passed on from one generation to the next and the latter because it is seen, literally, to dwell in the charismatic individual. If we pass from the more obvious examples of charisma associated with spiritual or religious movements to consider political charisma, then an appeal by a political figure (whether the fascist Oswald Mosley in Britain or Nelson Mandela in South Africa) based on personal magnetism or the presentation of extraordinary gifts may be considered as an act of power in and of itself.

---

6 The assumed effectiveness of authority as a mode of power stems from its *de jure* quality, where the mere assertion of authority is deemed rightful. The reified nature of this claim goes some of the way to explain the conflation of such roles with power. In contrast, *de facto* authority directs attention to the exercise of power and the experience of being on the 'receiving end' of such a relationship. See R. S. Peters (1967) on the formal nature of the distinction.

For Weber, what makes this appeal an act of charismatic authority is the recognition given to it by others on the basis of certain qualities and traits imputed to such figures. There is no organizational position attached to charismatic authority lending support or prescription to the role, only the act pure and simple. The appeal remains authoritative for as long as people are willing to submit themselves to the will of the charismatic leader. The relationship is the tie that binds. Should allegiance be withdrawn, however, should the authority of the figure in question be called into doubt, then the effect is weakened and their power diminished accordingly: the interpersonal relationship dissolves and the charismatic effect simply evaporates.

Of course, in the everyday world of electoral politics events are rarely, if ever, so one-dimensional. Other possibilities present themselves to political figures to exercise power over others. Lipovetsky (1994), for one, recognized that contemporary politics in western-style democracies, for example, adapted itself fairly rapidly to the development of image making and political marketing:

> In the United States, since the 1950s, political communication has come increasingly to resemble modern advertising; it uses the same principles, the same techniques, the same experts. Electoral campaigns are orchestrated by publicity agents and media advisors; one-minute spots modelled on commercials are produced; methods derived from motivational research are applied to speech writing and image building on behalf of political figures. After commercial marketing comes political marketing: it is no longer a matter of ideological conversion of the citizenry, but of selling a product in the best possible package. The austere heavy-handedness of propaganda has given way to the seduction of contact, simplicity, and sincerity; prophetic incantations have given way to the enticements of personalized shows and the transformation of political leaders into stars. Politics has shifted to a new register; it has been annexed in large measure by seduction. (1994: 168)

On this view, political *seduction* has superseded any broad sense in which crusading political leaders may have been considered charismatic and, perhaps more tellingly, it has displaced long-standing styles of authority based upon paternalistic attitudes and ritual spectacles. This shift in political styles receives a qualified welcome from Lipovetsky, principally for its less authoritarian guise and concomitantly for its more populist appeal, which leaves those in the political limelight more vulnerable and less detached than hitherto. The more accessible nature of this political style, the flawed character of the major performers, is thought by Lipovetsky to give modern-day politics its more modest appeal: power, as he says, 'has come down from its pedestal' and left political consumers with a more undifferentiated landscape in which to make up their minds. Patronizing views espoused by pre-eminent authority figures or impassioned speeches designed to bludgeon voters have given way to a politics of seduction, where politicians (Tony Blair for one) seek to take advantage of existing attitudes and values and turn them to their advantage.

Although this itself is seductive as a description of the contemporary European political scene in relation to personality politics, it is important to maintain the fact that seduction and authority should not be seen as alternatives in any power arrangement. Instrumental arrangements of power are not composed of one mode of power as if, in the above context, authority has had its day and events have now moved us on to the realities of seduction as a form of political control. Nothing could be further from the truth. Acts of authority and seduction may well play across one another in such a political arrangement; or acts of authority may mutate into outright coercion should recognition be withdrawn; or such acts may overlap with the effects of institutional domination as options are narrowed down and fixed to eliminate political choice. Seduction as a peculiarly indeterminate mode of power may even be eclipsed by a more determinate form. And so on. It depends, as ever, on the event in question and the agencies involved, although the one constant is that power most often makes its presence felt through a variety of modes playing across each other.

Consider, for instance, the attempt in the late 1990s by a small number of then little-known biotechnology firms to introduce genetically modified materials into the European food supply chain. In a short period of time, the companies found themselves facing concerted opposition to the arrival of genetically engineered foods, especially in the UK, principally although not entirely because of the backlash from the recent BSE (mad cow disease) scare. The big producers of genetically modified materials, the North American company Monsanto, in particular, were clearly taken by surprise at the scale and directness of the campaign waged against them. Their response (documented in Allen 2000a), and indeed the response of the UK government, provides an illustration of the power arrangements involved.

Exasperated by the inability of the consuming public at large to perceive the 'obvious' merits of genetically engineered foods and their general, all-round safety, Monsanto launched in the spring of 1998 a million-pound advertising campaign to win the 'hearts and minds' of the British consuming public. Full-page advertisements were taken out in the national weekend press extolling the virtues and benefits of genetically engineered food. What was noticeable about the appeal was its measured tone, designed to engage with the seriousness of the issues involved and to demonstrate an awareness of alternative views on the subject of genetic engineering and food. As an attempt to seduce consumers into eating foods that they might otherwise not have done, the appeal also drew support from two 'authority' sources: one traditional, which pointed to the place of the new seed technologies in a long line of traditional cross-fertilization practices, and the other more rational in style, based on 'scientific' evidence which can only be gained by access to specialized bodies of knowledge and expertise.

At the same time, Monsanto was busy attempting to acquire companies further along the food supply chain, in particular seed companies which, because of their pivotal position in the distribution networks, would provide a big gene technology producer with ready access to global markets. Farmers too, it was alleged, were being coerced into restrictive corporate contracts which left them with no choice but to purchase fresh seed from the same source year after year. If a biotech company managed to manoeuvre itself into a near monopoly position across a range of geographical markets in this way, the effect would be to take them closer to domination.

Thus, from one event it is possible to trace how an act of seduction was overlain by 'voices' of authority and ran alongside possibilities for both domination and coercion. With hindsight, the institutional strategies appear somewhat neater and more deliberate than was actually the case, but what we can take from this is the diverse web of relationships between the different modes, which amount to, as Foucault would say, an intentional but non-subjective arrangement.[7]

The attempt by Monsanto and other biotechnology producers to bend the will of consumers was far more ad hoc and opportunistic than perhaps the strategists of these corporations were prepared to admit. The fact that power was exercised with a particular end in mind sits alongside the fact that with every shift in public opinion, with every failure to win over public confidence to genetically modified foods, the institutional response was thrown further into a state of uncertainty and reflex reaction. Moreover, the response of the UK government to the volatile nature of the situation was equally ad hoc. Surprised initially by the strength of consumer rejection, the government's immediate reaction was to rely upon the authority of science to back up claims as to the safety of genetically modified foods, only

---

7 'Power relations are both intentional and nonsubjective. If in fact they are intelligible, this is not because they are the effect of another instance that "explains" them, but rather because they are imbued, through and through, with calculation: there is no power that is exercised without a series of aims and objectives. But this does not mean that it results from the choice or decision of an individual subject; let us not look for the headquarters that presides over its rationality; neither the caste which governs, nor the groups which control the state apparatus, nor those who make the most important economic decisions direct the entire network of power that functions in a society (and makes *it* function); the rationality of power is characterized by tactics that are often quite explicit at the restricted level where they are inscribed (the local cynicism of power), tactics which, becoming connected to one another, attracting and propagating one another, but finding their base of support and their condition elsewhere, end by forming comprehensive systems' (Foucault 1984: 94–5).

to find themselves faced with deep hostility and scepticism instead of the anticipated public deferral to 'expert' knowledge. Not long after, the government resorted to a strategy of inducement to gain acceptance of genetically engineered foodstuffs, once it was clear that the authority of their scientists had passed unrecognized.

The key point is that such mutating relationships are not exceptional; they emerge in this fashion precisely because they change in response to new and unforeseen circumstances. Yet they may remain intelligible to the eye, whether or not they are misguided or ineffectual. The contradictions are real, not merely apparent. Which, incidentally, is one reason why any coherence attached to a programme of multiple practices and techniques in the name of 'government' should be treated with a measure of suspicion. Instrumental arrangements of power are neither divisible in that way nor as effective as they seem.

Associational power, or power with others

Up to now we have considered power almost exclusively as an instrumental arrangement, as a series of actions designed to make others act in ways which they otherwise would not have done. There is another way of thinking about power, however, which has little to do with bending the will of others and rather more to do with the formation of a common will. The roots of this collective notion of power were discussed in chapter 3 and seen to reach back to an idea of power as a means to get things done or facilitate action. Perhaps the key aspect to this understanding of power is that it does not reduce itself to a zero-sum game where the benefits are divided unequally, but rather lends itself to a positive-sum scenario where all may benefit. Before we fall too quickly under the positive spell of this scenario, we should recall that the collective efforts of groups or organizations may just as easily lead to division and domination as to ends of a more instrumental nature. Having said that, to recognize that collective, integrative action can lead to the exercise of power with rather than over others is to take a step in an altogether different direction: namely, towards associational power. It is this direction that Arendt did much to open up.

Association, for Arendt, is about the bonds that hold people together when they are in pursuit of common, agreed ends. In one sense it is all about the benefits that may be reaped whenever people relate to one another in public life, and the quality of those relationships; it is about the power to connect, to bring together but not to suppress the interests and differences that commonly divide. Without being too sceptical about the extent of such arrangements, what we can take from this interpretation is that power, clearly, is perceived as a force for good. It entails a positive, not a negative, relationship, where mutuality, the prospect of an integrated 'we', represents an end in itself.[8]

The positive side to power was also something that was pointed up by Foucault, as noted in the preceding chapter, although the emphasis he placed upon the exercise of power closing down rather than proliferating possibilities for individuals should not pass unremarked. Both Foucault and Arendt, however, recognized the enabling condition of power which could lead to both self-development and the prospect of empowerment. The latter, in particular, when considered (by Arendt at least) as one of the outcomes of mutual interaction, also highlights the flawed understanding of association as merely an act of resistance by the powerless. In Arendt's view, people are not powerless; they exercise power not in a negative way, against a 'powerful' oppressor, but through the positive strengths of collaborative association. So in this kind of collective power arrangement, what exactly is exercised?

Perhaps the key to what we are looking for comes from Arendt's concern with what may be called *transverse* modalities of power; that is, those which literally cut across the many vertical power relationships maintained in conventional organizations or fought over in distributional struggles. Earlier, in chapter 3, I spoke about the kinds of association put together by those involved in certain human rights campaigns, or the kinds of collaboration which have been the hallmark of certain environmental protests, as examples of mobilization designed to overcome sectional interests by organizing around issues faced in common. If we extend the moral sentiments and political energies bound up in these examples of public mobilization to the collective opposition to genetically modified foods in the UK discussed above, it is possible to identify lateral relationships of power being exercised to fix a collective orientation. The two which come readily to mind are those of negotiation and persuasion.

---

8 Arendt's activist conception of power was outlined in chapter 3 and related to networked conceptions of power based, in this instance, upon mutual action. It is a matter of debate, however, how far Arendt is able to preserve differences of colour, gender, ethnicity, sexuality, class or age under her conception of 'plurality'. The identification of power with public space and the association of many wills and perspectives outlined in *The Human Condition* (1958) tends to gloss over the fact of social differences and their structural inequality. Her concept of 'natality', the notion of being-in-the-world-with others, essentially defines a human realm premised upon communication and action that speaks to differences of interest and opinion rather than subject position. Having said that, difference through equality, the recognition of the potential variety of alliances and their short term, intense nature, is compatible with Arendt's idea of temporary unity around an agreed purpose. See Disch (1994) on the notion of power as promise-making, and Benhabib (1996) on the role of Heidegger's work in the formation of *The Human Condition* and many of its central concepts.

Rather than think in terms of resistance, rejection or plain hostility to the introduction of genetically engineered foods in the UK, it is perhaps more accurate to highlight the strength of alliances between consumers, environmental activists, organic farmers, and such bed-fellows as the Soil Association, the Royal Society for the Protection of Birds, Christian Aid and Oxfam. In order to press a universal claim, these particular groups attempted to negotiate a common platform and stance among the many wills involved.

Negotiation in this context took place between disparate groups with a marked difference in the resources at their disposal, yet it required no obligation to comply on the part of all those involved. As a communicative model of interaction much in the line of thinking developed by Jürgen Habermas, negotiations were directed at reaching agreement over common ends, not the foisting of one group's views on the others.[9] With little possibility of seriously disrupting the overall operation of the biotechnology giants, negotiation between the different groups was an act designed to demonstrate the 'powers' (read 'resources and abilities') of association. Co-operation between such unequal partners may be maintained on this basis for as long as the negotiated momentum is sustained. The fact that certain environmental activist groups also opted for direct action, destroying dozens of genetically modified crop trials at test sites throughout the UK, does not detract from the positive-sum gains of negotiation. It merely reaffirms the point that transverse modes of action may co-exist with those of a more instrumental character in any particular arrangement.

Persuasion too has its place in such arrangements. In a similar vein to negotiation, acts of persuasion are effective only in an atmosphere of reciprocity where all parties are prepared to listen and communicate. Arendt asserts that persuasion, in contrast to authority, 'presupposes equality and works through a process of argumentation. Where arguments are used, authority is left in abeyance' (1961: 93). Which, if accepted, implies that persuasion is more likely to be exercised effectively among equals, whereas negotiation requires no such conditions. Both negotiation and persuasion, however, are distinguished from modalities of a similar guise, such as seduction or inducement, by the symmetry of the relationships involved. Where

---

9 Habermas's theory of communicative action and the centrality of the public sphere as indicated in chapter 3 owes much to Arendt's ideas as developed in *The Human Condition*. See *The Structural Transformation of the Public Sphere* (1989) by Habermas and his critique of Arendt's classicist conception of power in a special issue of *Social Research* (1977) devoted to the ideas of Arendt and their influence. The two differ in many respects, however, not least over their interpretation of the nature and role of the public sphere as the site of mobilization and the basis of modern forms of political legitimacy. If anything, Arendt is too naïve over the possibility of negotiation achieving solidarity through mutual promises.

the former relationships require a two-way process of communication to exercise the 'power to' achieve shared outcomes, the latter relationships work through predetermined choices to exercise 'power over' those that have already had their interests aroused or their behaviour checked.

For some, to talk of reciprocity, two-way negotiations and shared outcomes rather than resistance and opposition is to take us out of the realm of power and into something like civics proper. Should it be thought that the modal basis on which power is exercised in collaborative arrangements is of a completely different order, however, one need look no further than Arendt's treatment of authority. Authority, she reminds us, can serve purposes other than that of bending wills instrumentally.

A good deal has been written about authority, much of it as we have seen of a hierarchical nature, but Arendt, in keeping with her keen sense of mutuality, chose to interpret the term laterally rather than vertically. More like advice than command or control, for Arendt, authority, as was acknowledged in chapter 3, is something that is exercised *among* rather than over people. This understanding is more classical than modern and refers to what perhaps rather quaintly we would describe today as an act of wisdom: people giving sound advice that it would be foolish to ignore.

Giddens (1994), as noted before, refers to such traditional figures as 'guardians', people who dispensed a form of wisdom gained through their position as the repository of all past knowledge. Today, such 'experts', as they would no doubt be called, are less likely, in Giddens's view, to be accorded the trust once due to them. Faced with a diverse range of opinions on almost all matters, the public at large has learnt to be sceptical about absolute 'truths' – as perhaps the controversy over genetically engineered foods or indeed that of genetic cloning in general demonstrates. Trust, in an age of grounded scepticism, on this account, is always revisable and may be withdrawn as casually as it has been bestowed; people make up their own minds, but do so by weighing up expert opinion rather than simply accepting it.

If this is so, however, it is just as plausible to entertain the idea that the 'authorities' which today have a place *among* the public are those which freely acknowledge their uncertainty in the face of something as complex as the consequences of gene technologies. To do otherwise is certainly to risk losing the trust of a large part of public opinion over the safety of such technologies. In that sense, it is possible to see how authority, negotiation and persuasion may play across one another in the formation of group alliances designed to realize shared aims, despite the groups' quite different backgrounds and interests. Like their instrumental counterpart, such arrangements are likely to develop in an ad hoc fashion in response to new developments, although their existence is perhaps even more precarious given the many wills and open agreements that may be involved.

Once we begin to think of power in this way, as distinctive in its modal composition, we can begin to see why *what* is exercised is of such crucial significance. To recognize that power and its effects are not reducible to any number of individualistic practices and techniques, collaborative or not, is to hold on to the understanding that there is something in and of itself to domination, or to authority, or to persuasion or to any other such modal form. To think otherwise is to miss the complex interplay of forces which give an event like the controversy over genetically modified foods in the UK its particularity as a composed power play (unscripted, of course).

Arrangements of Power

Much of this chapter has been of a ground-clearing nature, laying the foundations for power's topological qualities to be built up over the next two chapters. Some readers may have already convinced themselves that a concern with such traditionalists as Weber and Arendt is an unnecessary diversion, if not a waste of time then a rather prosaic attempt to set out the nature of power before moving on to the more interesting higher ground. Why not pick up from where the chapter started on the necessary spatiality of power and go straight to a more nuanced topology of power?

Well, it would have been possible to do so, but I think that there would have been certain risks attached; risks that could defeat the eventual aim of the exercise, which is to show the difference that spatiality makes to the workings of power. For unless we understand, first, that power is a *specific* kind of social relationship, not a general or ubiquitous one, much of what follows on thinking through the effects of power spatially would be of such a broad nature as to be little more than meaningless. In that sense, an old-fashioned concern with the different modalities through which power is exercised is, in my view, crucial if we are to avoid any preconceived notions that power as a kind of indiscriminate substance is all around us and works itself out rather straightforwardly in immanent fashion. In particular, an understanding of the different modalities of power, and what is distinctive about them, resists the simple equation that power equals resources. Having said that, it is equally important to preserve the sense in which resources of whatever kind, settled or otherwise, are part of any arrangement of power.

The kind of loose, articulated arrangements that I have in mind are in many ways a testament to the fallible politics of agency involved. Open and provisional in outcome, the effects of such arrangements, lest it be forgotten, are frequently played out in arenas where the unequal distribution of resources is evident to all concerned. While nothing is predetermined, whether it be a confrontation on the geopolitical stage between unequal participants or the interplay of forces within a welfare officer's waiting room, it would be naïve of me to ignore the fact that resources often do count. They can and do make a difference to the outcome of any arrangement of power. But, and this is the important point, we cannot and should not 'read off' the exercise of power from the resources available. If we do, we fall into the all too easy assumption that size matters, marshalled capabilities are all, and power is amassed on such a scale that opposition is futile. These are misleading assumptions which obscure rather than reveal the working of power to us, in whatever setting or place we happen to find ourselves.

To ask what power is and how such arrangements work is thus hopefully to reveal something of the nature of what we face or what we have to negotiate when confronted by forces bent on eroding choice or manipulating outcomes or striving towards more mutual ends. In that sense, the task of this chapter, as indicated at the beginning, has been to turn to new purposes certain previous thinking on power and space, or rather to initiate that process. If we have made positive use of Foucault's thinking about power as a relational effect coextensive with its field of study, yet turned it into something less nominalistic in its stance and meaning, I now wish to show precisely how geography is integral to power's mode of operation. In short, I want to demonstrate how spatiality makes a difference to the effects that power can have in a topology with an unfamiliar yet traditional twist.

6

Proximity and Reach: Were there Powers at a Distance before Latour?

In truth, the question is somewhat tongue-in-cheek, a direct reference to Bruno Latour's (1988, 1999a) argument that microbes did not exist before Louis Pasteur rendered them 'visible', although it is not without serious intent. The idea that power assumes extensive reach into the lives of those not present or draws distant others closer into its ambit is not one we should be complacent about. After all, you probably know what it means to have found yourself doing something which you did not really set out or want to do, although you cannot quite remember why, or indeed if anyone ever told you to do so. None the less, the experience of imposition, of not being entirely in control of your own actions, is real enough.

To that end, Latour was keen to render 'visible' what is involved exactly in the imposition of such constraints from afar. In particular, he was eager to show that it was possible to dominate others at a distance through the mobilization and translation of what he referred to as 'circulating traces'; that is, virtually any element, entity or piece of documentation that can be inscribed or re-presented in some way so as to hold the mobile world 'out there' constant. On this view, as we shall see, the ability to stabilize power relations at a distance rests upon the work of translation, in particular the inscription of a particular way of doing things so that its basic message remains the same regardless of whether those involved are near or far. Thus the movement back and forth, the successive translation of the world 'out there' towards the 'centre' and its translation back, as Latour would have it, is thought to bridge the gap between here and there. We are brought to order through a process of connecting, and the far-off is no longer quite as remote.

If what is meant by 'far-off' is just around the corner or at the other end of the country, then few would probably doubt the ability of, say, legal devices to have an impact at a distance, even if not always as prohibitive as imagined. Even so, we should still problematize what is meant by reach in this context and what kind of relational topology is involved. In fact, whether the places concerned in this dispersed landscape are far apart from one another, close by, or next to each other, I would still want to explore how power is mediated relationally: how domination, for instance, may take effect through successive interactions or establish an immediate presence through real-time technologies, or how seduction appears effective at arm's length whereas authority seems more telling in close proximity. In each case, I would want to establish *the diverse geographies of power's proximity and reach* and explore how they play across one another, often to bring the far-off within reach, but also to make the close-at-hand seem surprisingly distant.

This may be a different topology to the one that Latour appears to have in mind, although my quarrel is not with his more refined sense of spatiality, as will become evident. If, however, as I argued in chapter 5, power is always already spatial, then we are obliged to think through the effects of power spatially, rather than dwell upon its spatial aspects. What I want to do in this chapter is to develop a further stage in this argument, by first taking the distances that Latour seems to have in mind and problematizing his implied notion of spatial reach. And then, because of its obvious debt to Latour, I work through a Foucaldian-inspired account of how power, or rather government, is said to operate at a distance, to show why exactly the account makes little sense unless it is refigured to bring out the modal qualities of power in terms of their proximity and reach. Finally, I shift to a 'global' setting to make a parallel argument in relation to those who remain enamoured of the simultaneous reach, control and buzz of electronic networks to bridge the gap between here and there.

Powers of Reach

To be fair to Latour, he does point out that the ability to act at a distance by drawing others into a network of connections designed to realize the 'will' of those holding the arrangement together is not always to do with power. Acting at a distance on people, events and locations does not necessarily involve relations of power, instrumental or otherwise. Actions can simply be that: actions with or without certain purposes in mind. However, what is useful about Latour's formulation as to how to reach across distances is that it draws attention to the *mediated* action involved in sustaining relationships at arm's length, where the ties that bind are both mobile and open to reinterpretation. If, for much of the analysis, the notion of power entertained is, arguably, rather anaemic, that does not detract from the force of the argument that nothing, least of all power, merely travels or radiates out intact from a central location to produce a predetermined effect elsewhere on the networked landscape.

In one sense, Latour (1986, 1987, 1999a) has helped to render visible something of what is involved in establishing and maintaining ordered lines of conduct at a distance. This in itself is a kind of explanatory leap, although as we shall see, where power relationships are explicitly involved, the ensuing arrangements he describes are surprisingly rigid and closed. More to the point, it is only one description of how distances may be overcome or how power is able to reach out to enmesh those in dispersed locations. If, as in many a past colonial arrangement of power, distances were overcome in Latour-like fashion, as a *succession* of mediated relationships, each, to acknowledge Foucault, with its own spacing and timing, in contrast, the reach of power in a variety of contemporary electronic settings might be just as likely to be characterized by its *simultaneity*. Rather than overcome distance by traversing it, the gap between here and there may also be bridged by compressing space by time. Distance, on this account, is measured more by time than kilometres and brings with it the experience of a more immediate connection to those absent 'elsewhere'.

Before we consider the possibilities involved, not all of which are recent by any means, it is perhaps best to follow through, first, precisely what is entailed by the idea of power's successive reach.

Holding the world at arm's length...

For Latour, the ability to hold the mobile world constant when much of what is 'out there' is absent or far away from those seeking to order it comes down to an ability, as mentioned before, to mobilize and translate the 'worlds' of others. The ability to draw others into a particular colonial arrangement, for instance, and then to order their lives, in these terms, would have involved, as Latour intriguingly points out, 'somehow bringing home these events, places and people' (1987: 223). The idea that the British back in the eighteenth and nineteenth centuries 'brought home' parts of Africa, Australasia and India to London as the imperial centre, or that the Portuguese and the Spanish did likewise in the fifteenth and sixteenth centuries to much of the 'New World', is not as odd as it may at first appear, however.

Many things, as well as people, were made to circulate, when you consider the circumstances under which the colonial forays took place. Obviously there is the trade in human lives, but what Latour has in mind are all the objects collected, observations made and calculations performed by the colonial bodies, which, through their scientific expeditions, cartographic plottings and taxonomic zeal, were rendered fluid and mobile. In reaching out across distances in this way, the mobilization of artefacts, the gathering of information, and the recording and charting of absent worlds became the means through which administrators, scientists and politicians 'back home' extended or imposed not just their understanding on distant others, but their sense of order too. On this view, as long as the bundle of things mobilized – objects, knowledges, mappings and such – could be combined in some way so as to produce a stable, ordered arrangement, then the capacity to act at a distance was firmly in place.

The relevance of what Latour calls 'immutable and combinable mobiles' for stabilizing the networks of association which made possible the exercise of power at arm's length is by no means restricted to a past set of associations, however. Latour's general understanding as to what it takes to exercise power at a distance is precisely that: a generic framework which is applicable to any number of situations where power's extensive reach is at issue. Statistics, numbers, classifications, judgements, drawings, designs, records, files, in fact anything that can be mobilized and embedded in networks stretched over space and time, can be used in some way to administer people's lives at a distance – especially, it seems, if the trace takes a paper form:

> All these objects occupy the beginning and the end of a similar accumulation cycle; no matter whether they are far or near, infinitely big or small, infinitely old or young, they all end up at such scale that a few men or women can dominate them by sight; at one point or another, they all take the shape of a flat surface of paper that can be archived, pinned on a wall and combined with others; they all help to reverse the balance of forces between those who master and those who are mastered. (1987: 227)

To gauge the significance of this point, one need only think, for example, of certain contemporary government agencies whose task is to administer a uniform standard of service across a diverse and widespread population. Acting at a distance to deliver, say, a centralized standard of social welfare entails a wide range of inscription devices, from statistical averages and standardized assessments to universal rules and unvarying regulations, all of which are lodged in files and records scattered across multiple authorities, and translated by front-line officials who find themselves forced to draw upon their own discretion in making judgements. The central bodies in this scenario are said to maintain their pre-eminent position, however, by placing themselves precisely at the points where everything that flows back and forth (reports, contracts, service agreements, all kinds of paperwork), everything that is 'hooked up' to the welfare network (voluntary organizations, private trusts, commercial carers), has to pass through them in some shape or form.

In so doing, the well-placed bodies are said to be in a position to 'fix' a collective orientation through a complex process of constructed meanings and strategies for moving forward, which, to all concerned, appear to be indispensable and irreversible. Or as Callon and Latour put it, a particular configuration of power is stabilized by the enrolment of 'other wills by translating what they want and by reifying this translation in such a way that none of them can desire anything else any longer' (1981: 296).

The key to the success of this kind of arrangement, then, it would seem, is the ability to 'hook up' others to the process of circulation, to draw others into the network of meanings in such a way that it extends and reproduces itself through space and time. In some hands, the exercise appears to add up to little more than a process of resource mobilization of the kind outlined in chapter 5, where control over others at a distance merely represents the capacity to muster contacts, artefacts and information of one sort or another as a means to exert power (see for example, Law 1986a, 1986b). By retaining a clear focus on translation, however, Latour is able to show how the extension of networked arrangements involves more than the mere circulation of resources; rather it involves a mediated *exercise* of power, where distances are overcome by the *successive* enrolment of others to form something akin to a single will.

It is not, of course, suggested that establishing domination at a distance through the process of engaging, enrolling and obliging others to adhere to a particular line of action always results in a successful translation strategy. But it is thought possible that the more that the world 'out there' is made to resemble the kind of stable, ordered conditions produced within networks, the greater the likelihood of success. However this may be, it seems to be taken for granted that it is possible to replicate such controlled conditions, given the immutability of the many devices in circulation. Presumably in the case of centralized welfare provision, it is the standardized uniformity, the bureaucratic constants, which are thought to reproduce an environment where people feel able to write themselves into the scripts of the centre. Or in a colonial context, if the combination was right, if it was somehow possible to put into place all the convenient charts, tables, maps, plans, texts, instructions, ordinances and monitoring devices, and to make the arrangement picture-like, that is readily accessible to the imagination, then so much the greater were the chances of exercising long-distance control.

The persuasiveness of this view depends not only how much one believes in the actual immutability of what circulates between different settings at different times (is the meaning of such entities really so little dependent on context?), but also on the assumption that it is possible to replicate through translation strategies (in Latour's laboratory-like fashion?) the kind of ordered schemas beloved of those 'in control'. If, however, we take a different view, namely that much of what is extended across a network is not replicable in any simple way, that the arrangements are more open than closed, less connected than is hoped for, more prone to leakages than is presumed, then power's successive reach is rather more hit-and-miss than controlling. 'Something always escapes', as Deleuze and Guattari would have it (1988: 217).

On this reckoning, either much of the world remains outside of networks of one kind or another, despite attempts to spread the certainties of the 'centre' further out, or the networked associations are not as far-reaching or as durable as is sometimes imagined. Whether or not this is so is hard to say, given the case-by-case nature of the evidence offered in support of power's being effective at a distance. There are, however, one or two significant doubts to consider, which point to the more elusive, provisional nature of power when it is conceived as something which constitutes itself through the spacing and timing of the mediated relationships involved.

Chief among these doubts is the well-founded observation that any attempt to extend control over the varied lives of many individuals, whether through traditional forms of delegation or otherwise, may both enhance the control of the centre *and* lead to its distortion and dilution. Barry Barnes (1988) has drawn attention to this contradictory dimension to power where it is subject to extension. An element of discretion, he argues, is built into the very exercise of power: once authority is devolved, delegates are empowered and able to make independent use of their new-found capabilities. The dispersal of (positions of) authority in this manner thus opens up a greater possibility for ambiguity in the supervisory process and raises the prospect of displacement and reinterpretation.

Second, and in a related way, the wider the dispersal of power, the more opportunity there is at the many points of intersection with other bodies for agents to mobilize other resources, other sets of interests, and to shift the line of discretionary judgement in unanticipated and unforeseen ways, or even break with it. This possibility was remarked upon in the discussion of dispersed techniques of government in chapter 4, for example. While there is no necessary reason why the chain of meaning should shift decidedly from that produced at the 'centre' of a network, the larger the number of outside interests to negotiate, the more varied the mix of resources, the greater is the potential for the disruption and dilution of any far-reaching 'certainties'.

On both these counts, then, one needs to be cautious about claims which speak about the feasibility of domination at a distance based upon the replication and durability of mediated arrangements. If we leave aside all talk of the 'loss of power' that is said to accompany dispersal and concentrate instead on the degree to which people and places really are 'hooked up' to extended networks, then perhaps it is not so much distance which problematizes the reach of power as the series of mediated relationships through which power is successively composed and recomposed.

Having said that, a succession of mediated relationships through which arm's-length control may be more or less sustained over time is not the only way that distance may be overcome. Instead of *traversing* the gap between here and there on an incremental basis, the alternative is to *dissolve* it by establishing near instantaneous reach.

... or dissolving the far-off into the close-by

In a thoughtful piece on the changing relations between presence and absence as the direct outcome of a communications shift in the West, Rob Shields (1992) concluded that it was no longer possible to think of social proximity as something which always coincides with spatial proximity. In this, he was attempting to unpick the nature of presence in contemporary situations and draw attention to the fact that the distant world of, say, the far-off dealing rooms of some North American financial trading house or the seemingly remote seed markets in the poorer economies are now more or less *present* in the here and now of the western economic 'powerhouses'.

In lots of ways, this dissolution of the far-off, the lifting out of social relations from one context to another, as Giddens (1990) describes the process, has long been around. Stephen Kern (1983), among others, has documented this in, for instance, his observations on the altered sense of the 'present' which accompanied the shifts in news communication technologies at the end of the nineteenth century. Not only did the sense of the present 'thicken' for many in Europe at that time as their imaginations were stretched to encompass more than the local event, it also 'expanded spatially to create the vast, shared experience of simultaneity' (1983: 314) as more and more events around the world were drawn into the 'now' of everyday experience. And if it is valid to point out that distances were dissolving at that time as a result of rapid transformations in transport and communications technologies, many would now argue that the mediated simultaneity generated by contemporary technologies has heightened yet further the sense of immediacy through which we experience 'elsewhere'.

Be that as it may, it is certainly not necessary to endorse the oversimplistic and often reductionist visions sometimes attached to space-shrinking technologies to acknowledge that the spatial division of near and far has lost some of its poignancy as a result of the speed-up in communications. Which, of course, raises the prospect that far-off places and events may be 'brought home' instantaneously by the compression of space by time, rather than through an extended circuit of incremental actions at a distance. If to reach out successively at arm's length is one way of exercising power across space, drawing others within close reach through near instantaneous forms of communications represents an alternative means to hold the mobile world 'out there' constant.

Indeed, much has been made of the ability of the new telecommunication and media technologies to intervene and shape the lives of those absent in space and time. What were once conceived as the 'distant powers' of big political and economic authorities at both the national and international level are now, it would seem, capable of almost simultaneous effect at even greater distances. Notwithstanding the tendency to exaggerate the effectiveness of the mediums involved, as with the delegation and distribution of (positions of) power, the immediacy brought about by speed and mobility may both enhance the directness of power *and* lead to its dilution.

On the one hand, the new and not so new technological mediums, by virtue of their capabilities (for they are resource mediums) to recast relationships of proximity and co-presence, may enhance the exercise of power by reducing uncertainty associated with an extensive translation process and thereby serve to limit possible ambiguity and surplus meaning. On the other hand, the very same technologies, because they are designed with transmission in mind, risk a more superficial and indeterminate mode of application. With issues of transmission, transfer and co-ordination in the foreground, the nature of what is communicated, how it is interpreted and what it all actually means may receive far less or even scant consideration. Before we can evaluate the basis of this tension, however, it is useful to be clear about the kinds of mediated interactions that are under discussion.

Perhaps the first thing to point out is that the contrast between exercising power over space and time and exercising it synchronized into real time is *not* that one is mediated and the other not. Both ways of acting at a distance involve mediated interaction; it is just that the experience of space and time is qualitatively different in the two cases. Whereas the successive enrolment of others at a distance would very probably involve a mix of physically proximate and distanciated contacts, the spacing and timing of near instantaneous forms of communication take place between people who are present in time only. Their sense of 'nowness' does not entail spatial proximity; presence is experienced only through social proximity.

In neither way of acting at a distance, however, is it the case that power merely transcends space, as if all the relationships involved are less tangible than those which involve the body language and gestural symbols of co-presence. That may well be the impression gained through the language that is frequently used to describe the forms of mobility and movement involved in extending power's reach, but in practice power is constituted directly through co-present relations of an equally tangible nature. It is just that they are mediated in qualitatively different ways. If advances in telecommunications, for instance, facilitate simultaneous exchange at ever greater distances, then that is not because power seemingly moves faster;

it is because resources are available which are capable of producing direct mediated action to bridge the gap between near and far in real time. In so doing, however, the medium itself can make a difference to what effects the exercise of power at a distance may actually have.

That is certainly the view of John Tomlinson (1999), who argues that there is more to inter-mediated technologies than simply the ability to deliver communications across tracts of space and time. In his view, the means of communication, whether the telephone, the computer, the television or some other electronic medium, are far from being empty vessels. Rather, their technical limits and material form hold particular consequences for how the different mediums are used and experienced socially and, following John Thompson (1995), Tomlinson isolates three modes of communication, each one characterized by its own distinctive form of interaction. Without going into detail, the first refers to direct, face-to-face forms of interaction; the second to the forms of mediated technical interaction associated with the telephone, the internet, e-mail, fax modems, voice mail and letter mail; and the third to the one-sided forms of interaction (if that is indeed what they are) identified with print, radio and televisual broadcasting. Each mode is said to possess characteristics which shape the nature of the interactions involved and, although he does not draw out the point, each has some bearing upon how different kinds of power relations may work themselves out at a distance.

In discussions of this kind, face-to-face interaction tends to form the benchmark against which all other forms of mediated interaction are judged, sometimes justifiably. Thompson is not alone, for instance, in pointing to the lack of contextual clues associated with many forms of electronic communication (the absence of facial expression or voice intonation, for example) which may increase levels of ambiguity and misinterpretation. Despite their essentially dialogical character, there is, as Deirdre Boden and Harvey Molotch (1994) have stressed, much that can be overlooked or misjudged in telephone or e-mail conversations, where emotions are screened and sincerity impossible to gauge. A preoccupation with the organizational and transmission capabilities of these technical mediums can lead to an overestimation of their effectiveness and, more importantly, in relation to issues of control, delegation and instruction, conflate their installation with their effect.

As with the networks of long-distance transmission spoken about before, therefore, there are doubts attached to the effectiveness of drawing others within reach through instantaneous technologies. An element of discretion and an independent use of power remain a constant possibility, although the availability of real-time presence in the relationships may limit the scope of this. More significantly, the sense of presence projected on to the new technologies in particular does not, in and of itself, overcome the difficulties of long-distance communication. A web site may represent a point of presence on the internet, but it none the less remains the case that the pace at which information is shared is dependent upon how it is decoded and interpreted, not upon the speed at which it is transmitted. The fact that communications are fast, digital and electronic does not confer the properties of speed and mobility on their users or, indeed, guarantee rapid control.

Equally, the development of telematics and buzzing electronic spaces tells us little about the dispersal of power above and beyond the fact that a wider and greater number of locations may be within reach. For some forms of symbolic exchange, on the web for example, the nature of communications may be broad and indeterminate; for other forms of interaction, such as attempts by corporate headquarters to extend their authority through company intranets, the dominant presence may be precise and tightly focused. The simple point is that the effects of power do not stem from the technologies themselves; there is nothing about power that can be 'read off' from the instrumentation. Yet their material forms may under certain conditions make a difference to their impact. As always, it depends on the power relationships involved.

Likewise in relation to broadcasting media such as radio and television, where there are no equivalent points of application, the effective exercise of power has little to do with the development of ever more sophisticated cultural technologies. Where broadcasting differs from both face-to-face and telemediated forms of interaction, however, is that it is principally monological in character and directed towards a more diverse and widespread population. The technologies involved, whether based on print or on electronic media, are essentially mass media targeted at groups who may be indifferent to their intended effects. Although the nature of the contact may be characterized by a sense of simultaneity, the absent target groups or distant audiences remain outside the scope of detailed control, open more to the advances of seduction or persuasion than to anyone in authority.[1]

Thus, the great promise inscribed at the heart of all space-shrinking technologies, that the problem of distance may be overcome through

---

1 The exceptions here are those technologies which achieve a degree of immediacy through the use of electronic surveillance techniques to render visible the movement of others, usually in enclosed spaces. The often site-based nature of these disciplinary technologies – closed-circuit television cameras (CCTV), computerized tracking systems and geographical information systems, for example – monitor at a fine-grained level the activity of individuals at a distance. The precision and near instantaneous reach of these technologies distinguish them from other one-sided technological mediums, but along with their pervasive reach such forms of electronic surveillance share with the broadcasting technologies a rather shallow, indeterminate impact. As will be argued in chapter 7, seeking domination through extensive surveillance opportunities is rarely as intensive or as complete as often claimed (see also Graham 1998a, 1998b).

mediated co-presence of one kind or another, is rather vague as regards actual content, especially in relation to the effects of power. Which is not to say that the material form of the medium is irrelevant to the social outcome, for clearly the different technological mediums have some bearing on the possible forms of action at a distance. The prior question that needs to be addressed, however, is: *what implications do the diverse geographies of proximity and reach carry for the way in which particular modalities of power take effect?*

If, for instance, the acts of those in authority, as suggested in the previous chapter, are more telling in close spatial proximity, how potent is coercion at arm's length, given the highly visible and overt nature of the action? Does the short-term nature of coercive acts and the obvious hostility they generate imply they are less effective across well-defined distances? If a modest and rather suggestive form of power like seduction does lend itself to simultaneous effect at a distance, how compelling are acts of manipulation through similar technological mediums, given that concealment is part and parcel of the intended effect?

Government as Far-Reaching Authority!

Aside from Michael Mann, as was noted in chapter 3, it is rare to find theorists who consider a particular mode of power in relation to distance and what it takes to cross it. Whilst it is not uncommon to come across spatial treatments of domination, usually as a kind of predetermined effect extended over space, in practice domination invariably operates as a kind of shorthand for power anyway. For Mann, whose prime concern is with the organizational reach of institutional power, what is exercised seems, in my view, to take second place to issues of resource mobilization across extensive networks. So it is perhaps all the more novel to find someone who places authority directly at the centre of his analysis of how power operates at a distance. The body of work in question is that written by Nikolas Rose over a number of years and culminating in *Powers of Freedom*(1999). The book, which is subtitled 'reframing political thought', takes its substantive cue from the work of Foucault on government and, as such, was mentioned in passing in chapter 4. Here I want to focus more on the way in which authority as the centrepiece of power is taken up in Rose's nominalistic account of governing and, rather interestingly, placed at the core of what it means to govern at a distance.[2]

---

2   Rose's work is both prolific and influential and can be located within a growing trend which addresses the modern manifestation of government in such areas as insurance, health, welfare, poverty, employment and crime. See in particular the collection edited by Graham Burchell et al. (1991), *The Foucault Effect: Studies in Governmentality*, and the pages of the interdisciplinary journal *Economy and Society*. Of Rose's work, the most

The reference to governing 'at a distance' reflects the significance that Rose attaches to Latour's understanding of how action at a distance is made possible through mobile yet unvarying traces, or rather techniques, of government. Stressing the art of government in the liberal sense as an activity which intervenes to regulate the behaviour of widely dispersed populations, people, on this understanding, bring themselves to order through obligation and self-restraint. As we saw in chapter 4, the willingness of 'free' subjects to transform themselves in a certain direction holds the key to how control is exercised at points remote from their day-to-day existence. What appear to hold the whole apparatus together, however, and give recognition and due credence to the various inscriptions, pronouncements, edicts, judgements, aspirations and protocols in circulation, are the many and varied 'centralized' *authorities* in play. The extension of authority over particular zones of social activity, for example through the centralized standards and assessments practised by the welfare authorities illustrated earlier, is what enables, in Foucaldian terms, the governable spaces of the family, the workplace, the school, the clinic and so forth to take shape. Or in Rose's words:

> Liberal rule is inextricably bound to the activities and calculations of a proliferation of independent authorities – philanthropists, doctors, hygienists, managers, planners, parents, and social workers. It is dependent upon the political authorization of the authority of these authorities, upon the forging of alignments between political aims and the strategies of experts, and upon establishing relays between the calculations of authorities and the aspirations of free citizens. I describe their mode of operation as *government at a distance*. Political forces instrumentalize forms of authority other than those of 'the state' in order to 'govern at a distance' in both constitutional and spatial senses – distanced constitutionally, in that they operate through the decisions and endeavours of non-political modes of authority; distanced spatially, in that these technologies of government link a multitude of experts in distant sites to the calculations of those at the centre – hence government operates through opening lines of force across a territory spanning space and time. (1999: 49–50)

All this distancing by independent authorities and experts, however, is not something that, once set in train, works itself out in accordance with a kind of

---

obvious publications that locate authority as the centrepiece of government are Rose (1992, 1993, 1994, 1996a) and the related Osborne and Rose (1999, discussed in chapter 4), although many of the ideas expressed are developed in *Powers of Freedom*. For a different angle on what it means to govern at a distance today, see Davina Cooper's *Governing Out of Order: Space, Law and the Politics of Belonging* (1998), which introduces the notion of governmental excess and the limits, or rather precariousness, of institutionalized spatial authority.

immutable logic drawn up at a distant 'centre'. Rose is only too aware of the role of 'translation' in forging loose and flexible alliances between experts of different hues and colours, so that any particular objectives, say for the yardstick used to calculate efficiency in social welfare programmes or the definition of value for money in healthcare services, would necessarily involve attempts to broker an understanding which experts and skilled professionals alike could subscribe to without loss of face or judgement. Borrowing the language of enrolment and obligation from Latour, Rose is disconcertingly clear about how the process of governing at a distance operates.

Basically, through a process of mobilization, the truth claims of a range of accredited authority figures, under the guise of neutrality and efficiency, set out the norms of conduct that enable distant events and people to be governed at arm's length. Naturally, the fragile and rather unpredictable nature of the translation process can prove problematic. But Rose evidently believes that the ability of the experts to render thinkable and, in turn, promote a particular way of being generally holds sway. So, for example, in relation to social welfare, the broad claim that it is right to 'take responsibility for our own lives', or to respond to a 'hand up' not a 'hand out' from the 'caring' state, becomes a lived 'truth', endorsed by health professionals and other experts as intuitively obvious and supported by a variety of reports, surveys, statistics and funding calculations which inscribe meaning in a more stable, durable manner. In other words, the welfare authorities, in this case, and their practices of inscription make it difficult for people to think and do otherwise. Or so it would appear.

Laboratory-like authority

One of the consequences of borrowing ideas and slotting them in to a preexisting formulation (Foucaldian in this instance) is the excess baggage that accompanies them. The inheritance in this case is the laboratory-like conditions required for a successful translation strategy to obtain; namely that the 'outside world' of daily life be given the same paper form as that produced within the extended network and its 'authoritative' centre. Bureaucracy, it appears, with its Weberian attention to rational decision-making, impersonal authority and stable documentation, is favoured by Latour for precisely those reasons: it is thought to deliver a world 'out there' which closely resembles the certainties and rigidities executed at central office (see Latour 1987: 254–57). In these sorts of scenarios, where obedience is secured from above by those *in* authority, it is easy to slip into the kind of seamless logic that implies that, because various programmes, inscription devices or whatever are in place, the process of ordering will continue apace the longer and the more durable the governing networks.

It may be true that Rose considers his 'networked' account of government to be open, more or less connected and provisional in outcome – all of the weakly developed points in Latour's appreciation – but it is hard to obtain that impression from the overall tone of *Powers of Freedom*. Perhaps because he draws much of his thinking from early exemplars of actor network theory, which were more prone to relentless accounts as to how order was achieved, his thinking exhibits a similar style of formulaic reasoning. Pat O'Malley (O'Malley et al. 1997; O'Malley 2000) for one, considers much of the government literature to be overly coherent and systematic in its formulation, devoid of the messy confrontation of political struggle and the irreconcilable differences between authority 'voices'. Indeed, it is relatively easy to gain the understanding from Rose's work that the conditions required to secure an alignment of meaning between experts, despite having to work across a complex of relays and interdependencies, are relatively unproblematic to establish. Certainly, there is no sense of the hit-and-miss treatment of power considered earlier, or in fact of the open-ended possibilities built into the translation process by Callon (1986) in particular. Even Latour's shift to a Deleuzian actant-rhizome vocabulary and what that signifies in terms of provisionality is missing from Rose's relentless formulation.[3]

But perhaps this is all a little too fussy and what I should simply be drawing attention to is the absence of *relationality* from Rose's systematic deliberation. Indeed, one could be forgiven for thinking that no one was actually on the receiving end of the 'self-evident' truths expressed by the authorities, so little do we hear from the recipients. The extension of authority over a multitude of dispersed wills seems to be judged more by its intended effects rather than its actual impact. Which is perhaps all the more surprising given that, according to Giddens (1994), we no longer live in a period where trust in *an* authority is easily given or where faith in the 'experts' is willingly conceded.

---

3 In John Law and John Hassard's edited collection, *Actor Network Theory and After* (1999), it is possible to gain a sense of the shift from the early formulations of actor network theory, many of which appeared closer to straightforward ordering frameworks than to the later, more provisional accounts of heterogeneous association and translation. On Latour's more open-ended style, see his 'On recalling ANT' in the above collection (1999b), his study of the failed Aramis transport system in France (1996) and a number of the essays in *Pandora's Hope* (1999a). Having said that, despite the more provisional, more complex topological landscape envisaged, the 'world-building' activities of actors still appear to be preoccupied with order-making of a successive and incremental kind. It is just the outcomes which are less certain. For some nuanced observations on the spatiality of actor network approaches, see Murdoch (1998), Hinchliffe (2000) and the *Society and Space* theme issue 'After networks' (Hetherington and Law 2000).

As was noted in the previous chapter in relation to the introduction of genetically modified foods, faced with a diverse range of opinions on almost any matter, the public at large have learnt to be sceptical about received 'truths', no matter how understated their package and presentation. Of course, people think for themselves and make up their own minds when considering the advice and guidance presented for their consideration, but, as Giddens goes on to point out, the grounded scepticism of the contemporary age implies that people's trust in experts is somewhat fickle, and may be retracted as swiftly as it has been given. Disenchantment, even with those who claim to be enabling others to think for themselves, is more a fact of life than an exceptional state of affairs.

What this all really indicates, however, is not so much that authority – even in its most refined and indirect forms – has lost its potency as that in the process of people bringing themselves to order, there is more going on in Rose's account of government and power than seemingly endless authority.

Seemingly endless authority

There is something rather odd about the claim that the whole weight of government, in both its state and non-state appearances, rests upon the exercise of authority, when seduction or manipulation or inducement, even coercion, would all seem to have a part to play in what it takes for people to bring themselves 'into line'. Equally, it seems unnecessarily restrictive for Rose to limit the agency of government to 'experts', through what seems to be his favoured injunction, the 'authority of expertise'. In fact, the elision of experts and authority is itself revealing. Experts, despite the scepticism with which their knowledge is likely to be held, are assumed by Rose to exercise authority on the basis of their professional skills, professed neutrality and therapeutic discourses. Why authority is the only mode of power exercised by experts or why it should be prioritized over and above all other acts of power remains something of a mystery. Just because experts are deemed central to the process of government does not mean that they never engage in the likes of manipulation or seduction. 'Experts' in the fields of social welfare, educational reform, family guidance or economic management may well be 'authority figures', but they do far more than exercise authority.

Those on the receiving end of the practices of government would surely bear out this observation, if only descriptively, but the significance of the point goes further than this. It is quite unrealistic to subsume under the guise of authority (or simply ignore) the many other forms of power that are exercised on a regular basis, for two reasons. One is that, as stressed in the last chapter and argued throughout the book, different modes of power produce their own specific effects; the qualities which mark out particular kinds of power relation do not all resemble the types of deferral and recognition that accompany acts of authority. Experts in whatever domain may well engage, for example, in all manner of seductive or persuasive acts to win people over which have absolutely nothing to do with people conceding authority to them. And the second is that the overreliance upon authority is unrealistic, because the kind of government at a distance that Rose describes at length could only really take shape on the basis of spatially varied arrangements of power. What Rose believes to be authority's effective reach over a multitude of dispersed wills may well in fact involve a successive combination of arm's-length seduction, indirect manipulation, extensive inducement and proximate styles of authority. Both points deserve elaboration.

On the first point, the subsumption of all forms of power under authority, Rose makes considerable play in *Powers of Freedom* of the role of expert authority across a wide range of political arenas – from colonial rule to active citizenship and from social welfare through to the enterprising self at the workplace and beyond. The common feature of authority's strategies, regardless of context, is that they work through power and freedom. People govern themselves, we are told, guided and directed by the ethos of experts and the way of telling expounded by the authorities that be. Yet if we glimpse only two of the areas mentioned, those which we have not yet considered, namely active citizenship and enterprise at work, it is not so hard to draw out the many and varied power relationships which overlap, run alongside or even replace one another in the constitution of such settings.

Perhaps it is not altogether surprising, given the emphasis placed upon self-regulation, that for Rose both the arenas of citizenship and work oscillate around notions of the enterprising self. Old forms of direct control at work and a sense of duty in the civic sphere no longer serve their purpose, and in their place are frameworks of expectations and assumptions that enable people to make themselves up in self-fulfilling ways. So, in the world of work, the organizational culture of firms serves to diffuse throughout the workplace standards of how people should act and think to be fulfilled in their job; managers exhort rather than formally direct workers to better themselves by setting attainable objectives, which are then assessed and rewarded when realized; and a 'voice' is given to workers through quality circles and the like so that they may, at one and the same time, add value to themselves and the organization. Those who fail to respond to the behavioural and attitudinal training or who fall short on basic commitment are screened out at various stages from selection onwards (see also du Gay 1996).

Likewise in the civic arena, people are empowered rather than duty bound to take responsibility for themselves, to construct their civic identity and lifestyle in line with the accountable self. Attitude surveys, opinion polls, focus group findings, the results of citizens' panels and other techniques are mobilized, inscribed and translated by civic and community experts who then claim to direct the way people should act. In the UK, for instance, the inculcation of responsible care for children within families through the practices of 'naming and shaming', the contractualization of pupil behaviour between parents and school, the programmes of community safety, policing and self-management, are all various means by which so-called authorities aim to bring the active citizen into being and thus into line.

That all of these various techniques and practices are more or less in operation in corporate and civic spheres is not really in question. Whilst their impact is largely unknown and underresearched, what is being exercised in terms of power relations is far more than authority. In Rose's Foucaldian approach (laced as it is with Latourian mechanics), we know an awful lot about the means through which power is exercised, the multitude of practices and techniques involved. Far less attention, however, is devoted to *what* power relationships are actually involved when people are exhorted to govern themselves.

If there has indeed been the kind of shift in government in the arenas of work and citizenship that Rose claims, much of what is practised does not accord with the label of authority. Take, for instance, the organizational culture of enterprising firms. If it is as saturating in its effects as we are led to believe, then it is more likely to work through acts of *inducement*, whereby employees are rewarded for their compliance in some way, rather than through recognition of the insight of management and other such figures (for example, management 'gurus' such as Peters and Waterman (1982) and their much-hyped charisma) held up as authorities. Over time, acts of inducement may coexist with something much closer to *coercion* within the corporate climate, as the monitoring, grading and assessment of workers secure their acquiescence through the threat of deselection from the programme of training, or, as it is more commonly known, dismissal.

Alternatively, an atmosphere of inducement and reward may mutate into something approximating an exercise in *manipulation* as the transparency of the corporate culture and its aims become a little too obvious to all involved. Faced with a lack of trust in what it means to be self-fulfilling at work, management may conceal their actual intentions regarding productivity from a workforce, or they may selectively restrict the kinds of information that they receive about what was suggested within quality teams or through informal feedback mechanisms. In so far as management are successful in manipulating employees, the simulation of choice and fulfilment may well flourish. In an atmosphere of mistrust and suspicion, however, no quantity of 'translated' figures and flow charts or customer satisfaction surveys and marketing aphorisms is likely to induce people to 'hook' themselves up to the network enterprise.

Should, however, an organizational culture be less than saturated with meaning and excellence, indeed quite modest in its entrepreneurial ambitions, then *seduction* rather than inducement or authority may form the prime mode of power in operation. Open, suggestive, seeking to motivate rather than calculate, seduction can be effective only on the basis that enticements may well meet with outright failure or rejection, even indifference. Standards and expectations about behaviour may still be diffused throughout an organization, ideals extolled by management and presumptions made by all and sundry about attitudes, but whether or not they motivate depends upon how far they are accepted by everyone, not on whether they are always put into practice. Acts of seduction do not aim to make over a person wholesale; they work through spontaneity and impulsiveness as much as anything, in the hope that it becomes difficult for people to think other than in the ways implied. Much of advertising works in this register, and so too do many of the messages and practices of enterprising organizational life and corporate culture.

In such an organizational milieu, the enabling condition of power may well fuse seduction with *negotiation*, where a two-way process of communication can lead to reciprocal benefits (so that value is really added to both parties). Without an obligation to comply, the relations of negotiation provide the basis for lateral forms of association which do not require either reward or the therapeutic 'translation' of authority to sustain them. Of course, in any particular corporate power arrangement, acts of seduction and the niceties of negotiation may be overlain or eclipsed by a culture where authority *is* conceded by others on the basis of a legitimate and recognized claim to manage. Though how far such relations *on their own* are able to secure a willingness to comply with the ethics of self-government or self-development remains hard to fathom.

A similar line of critique could be directed at the rationalities, programmes and self-regulation techniques which supposedly underpin the construction of active citizenship and the authority 'voices' upon which the whole edifice rests. If active, responsible citizens are actualized at all, it is not on the basis of authority relations alone. While an argument can be put forward for describing situations where people bring themselves into line without recourse to threats, violence, constraint or conflict as the result of the influence, opinion and testimony of authority figures busily translating what it takes to govern at a distance, it amounts to a painfully slim piece of reasoning. Neither seduction or inducement, nor manipulation or negotiation, rests upon conflictual or forceful constraint, so there is little ground for subsuming these different modalities under relations of authority.

Rose cannot be entirely held to account for this state of affairs, however, as the emptying out of conflict from the translation process and the running together of all kinds of power relations is characteristic of the ethnomethodological style of reasoning behind actor network theory. Another piece of excess baggage from Latour and Callon's work that has found its way into Rose's take on government at a distance is the ability to reel off seduction, persuasion, inducement and negotiation as just so many practices among others which make up the process whereby actors are mobilized and enrolled.

Lost within this bundle of practices, however, is not only the distinctive nature and effect of these different modalities of power, but also their inherent spatiality.

Far-reaching what exactly?

In talking about the formation of a new type of citizen or the appearance of a new type of enterprising employee, Rose obviously has in mind more than the odd remote constituency or distanced community. Government at a distance, if it is to mean anything at all, is assumed to address a wide and varied population 'out there' composed of a multitude of dispositions and free wills. The truth claims of authority figures, as we have had cause to note, are thus assumed to be extensive in their reach, diffused through a loose alliance of experts who are networked both to one another and to an 'authorizing' centre through which all translations must pass.

Now, it is not part of my argument to suggest that authority relations cannot possibly go the distance, so to speak. As stressed in previous chapters, there is no spatial template for power arrangements: acts of authority may be effective at any distance, successively or by establishing presence in real time. Having said that, if authority is to have the leverage over a widely dispersed set of wills that Rose clearly envisages, the internalization of what is and what is not appropriate behaviour does seem to require an extraordinarily high degree of recognition. Recognition, as you may recall, holds the key to the exercise of authority, which, in this context, is not about deference, but rather a willingness on the part of others to believe and not to find fault with the attitudes and sentiments expressed. As such, the claims must strike a chord with existing attitudes, interests and beliefs among sections of the population. The appeal remains authoritative, however, only as far as the basis of expertise continues to be acknowledged and doubts dispelled.

How far individuals in a variety of dispersed settings come to see themselves as active citizens through this means, therefore, is largely dependent upon the scope of recognition for ideas of self-management, community activism and the like. Yet at arm's length, in anything beyond the local civic arena, the impact of these programmes is likely to be superficial in terms of their hold on the imagination and confined largely to those whose interests and desires already dovetail with the attitudes of self-fulfilment expressed. This style of governing authority may on occasion possess sufficient reach, but the scope of the appeal and its intensity remain limited – no matter how many paper traces or other supposedly immutable devices circulate back and forth between scattered 'experts'. The assumption that authority integrates individual choice and the techniques of a responsible self into its own order may possibly work for those predisposed towards recognition, but its basic message is likely to pass unabsorbed by those who are not, especially if they are distant in both space and time. Besides which, it seems excessive to imagine that simply because certain ideas and values are disseminated their reception should be anything other than slight, given the sheer variety of often contradictory influences that people are subject to in their everyday lives.[4]

In fact, the kind of civic transformation that Rose appears to have in mind, certainly where it has taken root, is more likely to be the result of seductive or persuasive acts that touch the aspirations of individuals who are actually present. At best, such acts may shape the expectations of individuals, serving to captivate and to motivate people across a wide range of settings, but crucially with an acceptance that the ideas expressed and the beliefs expounded may remain fuzzy and only loosely assimilated. In place of diffused expertise about the responsible self, there is only hazy scepticism; in place of normative inculcation based upon attitude surveys and poll findings, there is only curious appeal. In this situation, the option *not* to participate is precisely what enables seduction, in particular, to be effective at a distance and the 'authority of expertise' less so.

By implication, what I am suggesting is that the authority–recognition relationship is less effective – not ineffective – as it is more complexly mediated, more distant. Certainly in comparison to seduction, or manipulation for that matter, neither of which emphasizes the internalization of

---

4   Deirdre Boden and Harvey Molotch (1994), in their seminal article 'The compulsion of proximity', stress the importance of co-present interaction for maintaining relationships of trust, commitment and detailed understanding. Whilst they overplay the importance of face-to-face interaction as an almost transparent form of mediation, pointing to what is 'lost' in translation, especially in technologically mediated forms of communication, there is a sense in which the more direct the presence, the more intense the impact of relationships such as trust, recognition and authority. See also Clive Barnett (1999), who argues that the more highly mediated and indirect the form of communication, where subjects are spatially and temporally absent from one another, the less intense, the more diluted the impact of rule tends to be. Like Rose's, Barnett's concern is with the government of a dispersed and highly diverse population.

norms, authority's constant need for recognition implies that the more direct the presence, the more intense the impact. In which case, proximity and presence have a significant part to play in the successful mediation of authority relations when confronted with a diverse and dispersed civic population.

Much the same, it seems to me, holds for the exercise of authority relations in corporate enterprises, and not just in respect of supervisory or control functions on the office or shop floor. Within a corporate enterprise culture, for instance, an attempt to project at a distance a shared vision through, say, a mission statement or a firm's charter, which embodies (inscribes even) the values of individual responsibility, commitment and accountability, across a series of widely dispersed workplaces is something of a tall order. Becoming responsible for yourself and your job too is not a message readily absorbed from 'on high'. In other words, a willingness to submit to such an 'authorized' goal is not likely to be achieved on the basis of remote exhortations. Such pronouncements from afar are likely to be less than convincing in their impact and best seen as an exercise in seduction, designed to win hearts and minds rather than to express authority.

Unless, that is, such acts are accompanied by authority in the form of local management. In which case, a face-to-face presence on site may enable the management to secure recognition both for themselves, as 'experts', and for the rather distant corporate set of goals. Drawing people into a discourse of enterprise, through training programmes, teamwork, collective initiatives and the like, is the means by which local managers may secure a willingness to comply to some of the more novel aspects of a shifting organizational culture. Proximity is telling in this respect, precisely because it enables trust relations to be formed as the workforce perceive that managers have already absorbed the meaning of enterprise and duly brought themselves into line.[5]

Inducement, too, has its place in such an arrangement and may operate from afar as long as those responsible in a corporate hierarchy are able to sustain commitment through the value of the benefits on offer. Over time, however, the rewards of both a self-fulfilling and a material kind may lose their attraction and with it the remote power of inducement. Likewise, the threat of dismissal for non-conformity to a new regime, which constitutes a

---

5 The exception is charismatic authority, where trust may be established at a distance but only for as long as the charismatic effect is maintained. Once recognition is called into doubt, as noted in chapter 5, the relationship dissolves and any authority along with it. Indeed for Weber (1978), the charismatic was the most precarious form of authority, in part because of the difficulty of sustaining the intensity of the relationship over time. See also du Gay (2000) on the charismatic quality of much contemporary managerial discourse, which aims to obligate people through their devotion to the corporate 'mission'.

coercive act, may be exercised at any point on a corporate hierarchy, although the continuous effort and resources needed to maintain and oversee the threat diminishes its effectiveness as a remote ploy. Threats lose their credibility if they are not followed through, and tend to fade if their presence is not continuously maintained. Alternatively, the whole operation may transform itself into a series of negotiations around ways of being at work that are performed on a day-to-day basis by dispersed management and workforce alike. The permutations are numerous.

The absence of a blueprint for arrangements of power makes it impossible to be anything other than speculative, although we can be more forthright about which acts of power are more effective at a distance and those which are less so. Thus in common with seduction as a mode of power, acts of manipulation are capable of quite extensive reach, because those subject to the latter may be simply unaware of the control exercised over them. The ability to mould the actions of others in a certain way without revealing directly the underlying motives, or indeed the absence of any need to interact at all, are characteristic features of manipulation. Concealment and indirectness are part of the very meaning of what it is to be manipulative; they distinguish manipulation as a mode of power, and it is precisely because of these distinguishing features that its exercise is effective at a distance. Manipulation has spatial reach.

So, for example, in relation to the pursuit of cultural excellence in corporate organizations, the flow of information and ideas – say through various training initiatives which speak about responsibility and accounting for oneself at work – may actually conceal a prior decision by a remote and distant management to reduce personnel and operating budgets by 'empowering' staff on the ground, so to speak. Thus profits and operating costs, not responsible selves, motivate management's behaviour, and training serves merely as a cover for other, less visible but more familiar, economic goals. The fact that power is exercised in one direction only gives manipulation its potential reach, although with the cost that if deception by management is suspected, total credibility may be lost and any advantages of distance reduced to a minimum.

The disturbing ease with which Rose is able to gloss over the many ways in which power is exercised and assume an undifferentiated spatiality is hardly unique. Given the influence of Foucault upon Rose's thinking, though, one would have anticipated that a sensitivity to the ways in which power produces its own field of organization would have been present. While Foucault would certainly have appreciated Rose's nominalistic treatment of power, the former would perhaps have found it odd that attention to the constitutive spatiality of power was sadly lacking. The excess baggage of Latour's laboratory-like authority has no doubt something to do with the confidence that Rose expresses in what appear to be systematic constructions of government and what they are capable of synchronizing at a distance. However, one would have thought that the sheer diversity of a widely dispersed population 'out there', whether at work, at home or in the civic community at large, and the many points of opportunity for power to mutate and distort, would have cautioned against an overly formulaic construction.

Once it is acknowledged that power is always already spatial, then it is possible to imagine a rather different topology of power; one sensitive to the diverse geographies of proximity and reach and the consequences that they have for the way in which government at a distance may take effect; a topology, for instance, where the need for authority to be established close at hand is played across the far-reaching effects of domination or manipulation in, say, the field of civic or corporate welfare to paint a more mediated, provisional picture of power.

It may seem unfair to isolate Rose's account and to dwell upon its shortcomings given the merits of his attempt to consider power *and* distance, but his formalism can all too easily dull the imagination. It can certainly lead to thinking that power in the shape of various practices and techniques can be effective at a distance in all kinds of asserted ways, when plainly that assumption is far from straightforward. Indeed, if Rose had been less tied formally to Latour's immutable practices, he might well have appreciated not only the nuances of power but also that a *succession* of 'authoritative' practices over space is only one way to reach out to a dispersed and disparate set of wills. The alternative is to establish an *immediate* presence through real-time technologies.

Domination in Real Time

Establishing an immediate presence in far-off places, or in nearby locations for that matter, through telemediated forms of interaction is another, rather different, way of keeping track of distant events in real time and potentially overcoming uncertainty. There are no guarantees to this process, of course, and the very speed which may synchronize people's lives may do so in ambiguous and imperfect ways. Even so, Rose's account would surely have benefited from a consideration of the possibilities that government synchronized in real time would have to offer – and what leverage is involved.

The presence of power, if it is possible to refer to it in that way, is none the less rarely established either successively or on the basis of simultaneous exchange in the kinds of arrangements mentioned. Rather, such arrangements, more often than not, are a judicious mixture of both. The kinds of government exercised at the workplace, for instance, or in the broader context of national civic reform, are likely to involve a succession of mediated relationships as much as those of a more simultaneous nature. Whether an influential presence is established through the communications networks of a widely scattered corporate enterprise or through the wavelengths of a broadcast medium to reach a dispersed civic culture, technologically mediated ways of exercising power are just as likely to be accompanied by more circuitous, face-to-face channels. Seduction, for one, may lend itself to simultaneous effect through various electronic mediums, yet it is unlikely to be exercised by either corporate or public agencies in isolation, if only because of its indeterminate, rather limited impact.

There are a number of ways in which this understanding of presence may prove useful in further unravelling the spatial constitution of power. Here I want to take the analytics 'global', for want of a better expression; not because the analysis requires a switch in scale from Rose's national arena to global government, but for the simple reason that the term 'global' provides an evocative setting for considering domination *in real time*. If one is pressed to think of the really far-off or the widely dispersed, then 'global' fulfils that requirement without really having to try. If nothing else, it provides a ready-made shorthand for eliciting so-called far-reaching powers.

Without wishing to indulge this line of thought too far, global reach and power are often to be found in the same sentence, usually without much thought having been given to their association. An exception, in the geographical literature at least, is the work of Saskia Sassen, who has been writing on electronic space and power since the appearance of her seminal text *The Global City* in 1991 (see Sassen 1995, 1996, 1998, 1999, 2000). Her work, which was briefly commented upon in the previous chapter, is interesting in this respect because it foregrounds the embedded practices of global control as found in the big financial houses, international law firms, sprawling management consultancies and global regulatory agencies, and explores their reach.[6]

---

6 Sassen is not alone in this respect, as other geographers, most notably Jonathan Beaverstock, Richard Smith and Peter Taylor (1999a, 1999b, 2000a, 2000b), have also endeavoured to explore the global reach of major corporate institutions from their respective city locations. Much of their work is concerned with establishing the linkages between global cities in respect of accountancy, advertising, financial and legal services. Where Sassen assumes the presence of such corporate activities in a number of global cities, Beaverstock et al. measure it, mainly in terms of direct investment and professional employment numbers. Their concern to establish a 'metageography' of global cities in terms of the relationships between them, however, appears to rest upon a scalar framework of power relations, where cities vie with nation states in a 'rescaling' or redistribution of capabilities of the kind mentioned in chapter 2. Sassen, in contrast, seems to have in mind an economic topology of power in which corporate bodies draw much of their ability from the simultaneity and immediacy of their networked connections.

I mention just two ways in which such bodies are said by Sassen to make their presence felt at a distance. The first envisages the new, instantaneous reach achieved by the many diverse specialists drawn together in places like New York, Tokyo or London, who are said to 'run' the global economy 'out there' by responding to the immediacy of the markets. The scenario outlined is again close to that described by Max Weber as one of domination, where formally free individuals on the other side of the world find themselves in a situation where they have little or no choice but to fall in line with the interests of the big global agencies. The difference from Weber's day, of course, is that this form of economic domination is now practised in real time. The second avenue of reach outlined by Sassen picks up on a different way that such agencies are seen to extend their influence across far-flung networks, and which comes closer to the more prescriptive, norm-inducing practices described by Rose; namely the normative standards and conventions laid down by 'global' law-making bodies and financial agencies. As before, however, we are led to believe that authority relations are all that is needed to achieve the desired effect. It is, of course, as we have seen, never that straightforward.

Let us concentrate upon the first of these avenues of reach as it broadly reveals the kind of predetermined thinking that often goes on where power is assumed to have a *simultaneous* presence in a variety of global settings.

A controlling presence?

Sassen is keen to argue that the sheer scale and complexity of global economic activities today are not something that is anarchic or out of control. It is not as if corporations, finance houses and those who broker the welter of transactions which make up the global economy go about their business in their own sweet way without the slightest hint of management or co-ordination. Rather the control of such activities, the orchestration and co-ordination of the fast and flowing nature of economic business, takes place in a few strategic locations which have the capacity – the skills, resources and expert knowledges – to do the job, so to speak. Banks of lawyers, accountants, financial analysts, management consultants and programmers capable of handling the everyday complexities of transnational transactions cluster in places such as New York, Tokyo and London and, in true Latour-like fashion, operate as a centre of calculation. Everything – the movement of business, money and ideas – passes through them in one shape or another in a way that enables these high-level service professionals to practise global control.

In contrast to Latour's grounded vision, however, these 'command points' in the organization of the world economy, as Sassen describes them, are embedded within virtual electronic networks which span the globe and effectively negate the gap between the economic 'here' and 'there'. The digitalization of economic activities, the increased pace of transactions and the immediacy of the markets have given rise to a sense of 'nowness' in the global economy that only few organizations are in a position to control. Among them, empowered by state-of-the-art telematics, according to Sassen, are the embedded networks of private finance:

> The financial markets, operating largely through private electronic networks, are a good instance of an alternative form of power. The three properties of electronic networks: speed, simultaneity, and interconnectivity have produced strikingly different outcomes in this case from those of the Internet. These properties have made possible orders of magnitude and concentration far surpassing anything we had ever seen in financial markets. The consequence has been that the global capital market now has the power to discipline national governments, as became evident with the Mexico 'crisis' of December 1994. We are seeing the formation of new power structures in electronic space, perhaps most clearly in the private networks of finance but also in other cases. (1998: 178)

Sassen goes on to argue that both the growing virtualization of economic activity and the centralized capabilities for global control are part of a 'new geography of power' which is steadily eroding the sovereignty of traditional state-centred systems of co-ordination.

How far one wishes to endorse the emergence of new power structures in electronic space is of course debatable, especially given that for much of the time Sassen conflates resources and power. But what concerns me here is the rather cavalier fashion in which control is 'read off' from the new-found technological capabilities of the global 'centres', and the assumption that the kind of economic power wielded by global service professionals has a presence that is both extensive and effective. From what has already been said, the ability of new technological mediums to recast relationships of proximity and co-presence cuts two ways: they may enhance the exercise of power by reducing uncertainty, or they may dilute this exercise by paying more attention to the transmission rather than the translation of what is communicated. It is one thing to suggest that in the fast new economic topography it is possible to have a presence world-wide in real time and quite another to say that this occurs with any real effect. The possibility of near instantaneous forms of communication tells us little or nothing about their impact, only the fact of their installation. The more extensive the network, the more varied the interests involved, the greater the number of wills to negotiate, the less likelihood there is of anything resembling a 'global capability for control'.

This is not to deny, of course, that the types of electronic network cited by Sassen are capable of establishing a presence of one kind or another through the amalgam of professional and managerial specialisms clustered in global cities. The critical issue, however, is not about presence *per se* but whether or not a *controlling* presence is established. Certainly, it is hard to avoid the impression that the far-reaching powers of global agencies are indeed capable of establishing a dominant presence across the dispersed economic networks. The specific nature of this domination is obviously thought to vary depending upon the character of the networks involved, whether market-driven commercial networks, government rule-based regulatory systems, or those internal to a corporate multinational's global operations, but the net result is much the same. If it is not quite the simplistic 'command over space' variety of control that is envisaged, those 'in control' are thought none the less to make their presence felt at a distance by restricting the economic possibilities of others so that they have little or no choice but to comply. This, effectively, is the *practice* of global domination (see Allen 1999b).

Or is it? Assuming that some form of power is necessary to promote the co-ordination of the global economy, no matter how provisional and uncertain, it is not so much extended in domineering fashion as coextensive with the 'global' field. Power constitutes the many economic networks, be they long or short, global or regional, through the spacing and timing of the mediated relationships exercised by an array of embedded agencies. More to the point, economic domination, as we have had cause to note, is rarely exercised in isolation by the likes of multinational corporations, finance houses and governmental agencies; it is almost invariably part of an institutional arrangement which mixes far-reaching with more proximate modes, circuitous with more immediate styles of power, to establish and maintain an influential presence world-wide. Indeed, much that is said to fall under the banner of global domination in economic terms is more likely to turn on a combination of power relations.

Consider, for instance Nigel Thrift's (1994) account of how the City of London maintains a global financial presence through a series of distanciated and face-to-face practices by representing itself as a *centre of 'cultural authority'*. In a style not dissimilar to Latour's account of how action at a distance is made possible through mobile yet unvarying traces, Thrift argues that the City's expertise in gathering, circulating and interpreting financial information has given it a pre-eminent position within the world's financial networks. This has manifested itself, Thrift suggests, in four related ways:

> *First*, the City is a nexus of *face-to-face* communication through which information is gathered and interpreted. *Second*, the City is a centre for *electronic*

information gathering and transmission. For example, by 1989 Reuters 'maintained 184,300 screens worldwide, providing groups of customers not only with instant access to information, but also allowing groups of them to communicate with each other, and so provide an international electronic market-place' (Michie, 1992, p. 185). *Third*, the City is a centre of *textual* interpretation, from the voluminous research analysts' reports circulated to particular potential clients, even to humble tip sheets. In particular, the City is now a centre for the *global financial press* – including *Euromoney*, *The Banker*, *The Economist* and the *Financial Times*. The *Financial Times* started a continental European edition in 1979, with a New York edition following in 1985. By 1993, 40 per cent of the paper's circulation was abroad.... *Fourth*, the City is increasingly home to many different global 'epistemic communities', occupational communities with their own specialized vocabularies, rhetorics, knowledges, practices and texts. From economists to foreign exchange dealers to Eurobond traders each of the communities tends to live in an increasingly specialized narrative world. (1994: 349–50)

As a site of authority, then, the City of London, on this view, has become a *recognized* centre of all things connected with markets and finance. The implication is that others, many absent in real time, are willing to give due credence to the advice and textual interpretations on offer and to operate from a position of trust – revisable should their options fail to deliver the anticipated benefits. This, then, is hardly a vision of global control, where a sense of pronounced 'nowness' brought about by the immediacy of the financial markets works to close down options. In fact, there is even more than financial experts strutting their authority going on here.

One of the points suggested earlier in respect of authority as a mode of power was that the more complexly mediated the relation, the less intense its impact, given authority's constant need for recognition. In a more open environment, where financial information and opinion are disseminated over a widely dispersed economic arena, authority's reception is more likely to be akin to the limited effects that characterize both seduction and persuasion. The ability to suggest a general mode of conduct, to exhort rather than prescribe, through the pages of the global financial press or over the internet, are closer to these more modest forms of power than to any set of authoritative voices, no matter how direct their presence.

The issue, however, as before, is not that seduction or persuasion have displaced authority as a means of maintaining the City's presence in far-off markets. Rather, in such a varied economic arrangement as the City's financial networks, the need for relationships of social and spatial proximity alongside those of a more distanciated nature reveals something of the varied reach and intensity of power relations. They are not all of the same spatial order. Thus if Sassen is correct to talk about the formation of new power structures in the electronic networks of private finance, she would have done well to consider the relationship between proximity and presence as it affects authority's need for recognition, the near blanket-monopoly requirements of domination if it is to be effective over a dispersed populous, and the modest qualities of seduction or persuasion that enable those relations to make their presence felt in real time despite their indeterminacy. In short, she would have done well to consider power's inherent spatiality, rather than just some of the spatial aspects of domination.[7]

Power at a Distance?

To be honest, I cannot help but believe that in both Sassen and Rose's thinking about how power works at a distance, some variation on Latour's 'laboratory-like centre' theme is present. Whether behind the scenes or not, for them, the ability to reach into the scattered lives of others seems to possess an almost systematic quality which, whilst mediated, is still capable of producing something that approximates to a stable, ordered arrangement. Obviously the traces of borrowed ideas are more in evidence in Rose's work, given Latour's acknowledged influence, but for Sassen there is the similar misplaced belief that the world 'out there' can be made to resemble the complexities and the certainties of the 'centre', be they such well-placed bodies as the strategic headquarters of corporate organizations or the agglomeration of service expertise that inhabits global cities.

This may be comforting, but it provides few clues as to what is actually involved in the exercise of power at a distance. One of the prime difficulties with the 'authoritative' centre view of power is that it tends to be judged by its *intended* rather than by its *actual* effects. Because various agencies within global cities have the skills and resources 'to run' the global economy, it is somehow assumed that this is indeed what happens. Power may not be transmitted intact from point to point in a loose Weberian sense, but the spread of certainties from the 'centre' outwards tends to be assumed and not evidenced.

---

7 It would of course be possible to go on to explore, in the fashion of Paul Virilio (1986, 1991, 1997, 2000), the 'here-and-now' forms of control that work through speed and simultaneity beyond the financial markets, by considering, for instance, the role of company intranets and corporate telecommunications in attempts by global headquarters to bring their scattered employees into line in the Weberian sense of command and prescription, but the point remains much the same. Speed itself tells us little about the effects of the new communications technologies, and time would be better spent disentangling how the relationship between proximity and presence affects power and its modal qualities.

The assumption stems, in my view, from a persistent failure to problematize the spatial reach of power. There are no 'blocs' of authority 'back home' waiting to be recognized, no ready-made strategies of domination on stand-by, only mediated relations of power which take a modal form, either reaching out to a dispersed population through a succession of relationships or drawing them closer by establishing an immediate presence through real-time technologies. As I see it, the relationship between proximity and presence plays across these two ways of bridging the gap between here and there – and it is this that problematizes the reach of power.[8]

In this topological landscape, fixed distances and well-defined proximities fail to convey how the far-off is brought into reach or the close-at-hand constructed at a distance. More to the point, power relations, I have argued, are constituted differently in this space-time landscape, in line with what distinguishes an act of crushing domination from that of seduction, or a manipulative act from the unnerving threat of coercion, for instance. Conceived in this manner, a spatial analytics of power, far from celebrating the problematization of reach, would treat it as merely the result of the ways in which provisional arrangements of power take shape. If Rose had arrived at a similar conclusion, he might also have understood what it takes for government to bridge the gap between here and there.

---

8 To be fair to Latour, this kind of topological landscape is closer to that outlined by Michel Serres in their joint *Conversations on Science, Culture and Time* (1995). A concern to loosen, or rather reassess, the rigidities of defined proximities and distances is evident in Serres's topological imagination, in contrast, as he says, to a 'metrical geometry' with its fixed co-ordinates (1995: 60–1). In Latour's hands, however, it seems to me that the weaving together of distant others into more or less connected, more or less long networks of association stresses only the *successive* dimension of time and space. The only times and remote places brought into the here and now appear to be those placed within a common frame of reference by successive translation (see Murdoch 1998; Bingham and Thrift 2000). The sense in which space-time may be 'folded' by near instantaneous forms of reach, and the relationship between proximity and presence disrupted, plays across the more familiar pathways which traverse the gap between here and there. I want to insist that it is this play across proximity and reach – involving both successive and simultaneous mediations – that comes closer to a loosening of defined distances and times.

7

Placing Power, or the Mischief Done by Thinking that Domination is Everywhere

Whereas the previous chapter was concerned to show power as intrinsically spatial, this chapter is an attempt to show spatiality as imbued with power. This may sound like a play on words, but the difference is significant, especially for understanding what it is like to be *placed* in the midst of a tangled web of power relationships. In the former, the focus was upon the difference that space makes to the way that power 'works'; here the focus switches to the interplay of forces which actively constructs the power-laden spatial circumstances we find ourselves within. This chapter thus has less to do with the diverse avenues of power's reach and rather more to do with what puts power *and* us in place.

Another way to highlight this switch in focus is to think of the previous chapter as largely about the spatial analytics of power, whereas this chapter goes further to reveal its synthetic credentials. Rather than dwell upon particular arrangements of power, we can now give fuller expression to the jumble of arrangements, the messy co-existences and awkward juxtapositions of power that characterize places. And in doing so, it is possible to trace such colliding spaces and arrangements to the ways in which people move in and around one another, sliding across one another's lives in a more or less powerful fashion. In short, it is possible to move towards a more phenomenological account of space which reveals something of the tangled web of power relations and their meaningful impact.

We know from the last chapter something about what it means to establish a far-reaching presence, but here I want to probe the reverse side of that relationship: what it means for different groups and institutions to establish a presence that makes the close-at-hand distant to others. The beginning of an answer to that question, I believe, lies in the ability of different agencies to make themselves felt in ways that give places their rhythmic complexity and ever shifting character, by establishing a presence in all kinds of powerful ways – a dominant presence, a manipulative presence, a seductive presence and so forth. It is this placement of forces that I wish to explore in this chapter, in part, once again, to reveal the modal qualities in play, but also to show how such forces give shape to the *cross-cutting* nature of social spaces.

The starting point is Henri Lefebvre's phenomenological vision in *The Production of Space* (1991a), which made clear that there was more to power and social space than a series of static distributions or ordered diagrams. More than most, he was able to show how the spaces of everyday life were imbued with power (see also Lefebvre 1991b). Though cast in a familiar domination/resistance mould, Lefebvre's radical humanism proved to be of great value in unravelling the rhythmic and routine construction of social space, despite its rather one-dimensional sense of power. As a means of avoiding the idea that domination may be just about everywhere, however, it is the *tangled* arrangements of power in customary places that I focus upon here.

Lefebvre's Smothered Spaces

The Production of Space is one of those books which, when it was finally published in English in 1991, seemed to hold out the promise of revealing all that we desperately wanted to know about space, but never quite knew how to ask. The attraction of the book, which bears the mark of an author who plainly thought editing a pointless activity, was both its generosity in enabling all interpretations to appear legitimate and its 'key-to-all-things-spatial' tag. Derek Gregory (1994) has done much to curb the excesses of early interpretations and claims, whilst Rob Shields (1999) has more recently attempted to settle some of Lefebvre's ideas and suggestions. For my purpose, and now that we are more familiar with the range of Lefebvre's spatial thinking, I want to focus upon the interplay of forces that lie behind the comings and goings of those who are said to dominate space and those who come to see themselves as part of someone else's space.

The contrasting experiences of social space that I have in mind, where difference, the traces of others, is effectively smothered by a more dominant *presence*, takes its cue from Lefebvre's more formal splicing of social space: namely, the distinction between representations of space and representational space. Without my wishing to dwell on the philosophical foundations which underpin them,[1] the former may be understood as the

---

1 In the extended introduction to *The Production of Space*, Lefebvre stresses the interconnections between spatial practices, representations of space and representational spaces. Indeed, he refers to the three moments of social space as a conceptual triad:

dominant spaces in any society, which today are perhaps best exemplified by the more formal, abstract representations that find their true expression in the rationality of the planned urban location or indeed any 'cold', uniform space that is devoid of personal feeling or real attachment. In contrast, representational spaces are regarded by Lefebvre as 'lived' spaces; spaces which take their shape literally from the daily routine of 'users' and their symbolic attachments. Thus, the very way in which particular spaces are imagined, for example how people make sense of their workplace or their local marketplace, also represents a means of living in those spaces.

More importantly for Lefebvre, the two forms of social space should be understood in relation to one another, with the potential for one to contradict or be in tension with the other. The idea of spatial contradictions, where the manner in which a space is routinely used by one social group may actually subvert or disrupt the dominant or controlling rhythm, conveys the sense in which the *same* place may actually play host to a variety of cross-cutting social spaces. If you think, for instance, of somewhere like the City of London, which in the last chapter was referred to as a site of cultural authority, such a description may appear decidedly one-sided or partial in its account when one acknowledges the presence of groups other than those drawn from banking and finance. True, what Nigel Thrift (1994) had in mind was the recognized expertise of those who gather, interpret and circulate financial information from their state-of-the-art offices in London's Square Mile, but such spaces are not confined solely to the activities of financial traders, dealers and their associates. There are other 'users', ranging from those who maintain the elaborate digital infrastructure and secure the steel and glass buildings to those who clean the many floors and cater to a culinary diverse population. Indeed, a veritable 'army' of 'other' service workers – tradespeople, messengers, cleaners, chefs, waiters, catering assistants, security guards, porters and the like – inhabit the same place and through their routine and often imaginative use of it may clash with the dominant interest.

the perceived, the conceived and the lived. This is a product of Lefebvre's particular 'take' on Hegel's dialectical logic, which in contrast to formal logic wishes to achieve a synthesis of thought and being (see Lefebvre 1968). The emphasis placed by Lefebvre, following Marx, upon practical activity led him to the view that a more libertarian organization of urban space should be sought on the basis of lived experience. The difficulty that this poses for an understanding of social space, however, is that it strips the formal representations of space (conceived space) of the 'everyday' spatial practices which produce and secure a dominant coding of place. In a related way, the categorical attachment of perception to spatial practices distorts our view of such practices as unmediated by concepts or at one remove from lived experience. For an elaboration of this view, see Allen and Pryke (1994).

The key issue to note in this respect, however, is that any potential clash of interests, symbolic or otherwise, may occur *within* the same physical space; that is, those who work, use and make sense of London's office space may find themselves jostling for position within it or carving out a particular niche for themselves amidst more dominant users. Bankers in the Square Mile, for example, may leave their imprint firmly on the City as a place of finance and in so doing mask the traces of others who routinely use the same space, but they do so by acting *in the midst* of others, not cordoned off or separate from them. They remain near to others, yet socially distant, as much a part of elsewhere as they are of the financial spaces of the City. In this way the impression of the City of London as a coherent place of finance may be achieved through the suppression of its many alternative users, although neither necessarily with total effectiveness, nor indeed without the creative and often unprompted use of space by others disrupting the overall rhythm.

In terms of power as something which is exercised rather than possessed, what is at issue for Lefebvre, then, is how social space comes to be dominated in this play of contradictions. How is it possible for certain groups to be able to *abstract out* the daily routines and practices of others? If the stress is placed upon the domination *of* space (rather than simply *over* space), then presumably this must involve the closing down of possibilities, the restriction of alternative uses, so that others have little choice but to acknowledge the construction of a singular space – even though they may imagine themselves moving within and around it.

So how does the domination of space take hold? In Lefebvre's case, it would seem principally to be about the ability to *represent* space in a particular way, to code it in a manner that suggests that only certain groups are present. It is this ability to smother difference, to suggest who should be seen and heard and who should not, that can give particular social spaces the impression of sameness rather than displacement and diversity. Above all, it is the vast array of spatial practices, from the routine walks and rhythms which endow a place with meaning to the coded gestures, styles and mannerisms which prescribe a certain use for it, that puts both us and power in place.

There are a number of implications of such a practised view of power. Spatial codes as the means by which a space is inscribed may be read formally and deciphered (as one reads an office plan or a town map), or they may be apprehended through sound (the animated buzz of an office, for example, or the near silence of a cathedral where, Lefebvre suggests, space is measured by the ear), or through a sense of smell (where, for instance, the incense-laden atmosphere of the cathedral tells you where you are and how to behave). Such codes involve an element of performance, the effectiveness of which will vary in line with the degree of competence achieved and the resources mobilized. But my aim here is not to pursue these practices for their own sake, as significant as they are for understanding how social space is actually produced. Rather my intention is to probe Lefebvre's sense of how space is claimed exclusively and to focus upon *what* exactly is exercised in the name of power.

Coding dominance

Staying with the example of London's financial district, Lefebvre offers an interesting insight into how monumental spaces, in this case those of banking and finance, generate a sense of their own 'membership'. Less concerned with the obvious material qualities, layout and size of buildings, Lefebvre wants to draw attention to what may and what may not take place within a particular space:

> The indispensable opposition between inside and outside, as indicated by thresholds, doors and frames, though often underestimated, simply does not suffice when it comes to defining monumental space. Such a space is determined by what may take place there, and consequently by what may not take place there (prescribed/proscribed, scene/obscene). (1991a: 224)

This takes us closer to the idea that certain buildings, notably the offices and trading floors in the City of London, amount to a collection of prohibited spaces whose use is carefully prescribed. Loosely understood, this would involve social and physical barriers that denied access to those deemed inappropriate. But Lefebvre seems to be pointing to something rather less bounded than that. The sense of membership that he invokes seems to have little to do with the formality of closed doors and rather more to do with who is *recognized* as present.

On this understanding, the prohibitive aspect of space reveals itself through practices of discrimination between the comings and goings of those who work alongside one another. Only certain groups are seen and heard on a regular basis, as the lines of visibility and waves of sound suggest unequivocally who is and who is not 'out of place'. Whatever sense of borders one wishes to entertain is, on this account, the outcome of social interaction and an explicit spatial coding.[2]

---

2 Georg Simmel, in particular, was also keen to stress that borders and boundaries were the result of social interaction and not merely finite barriers, even to the extent that he wished to argue that doors represented both separateness and the possibility of stepping out beyond them – a point clearly illustrated by the swing movement of doors. See his essay 'Bridge and door' (1997) and also Allen (2000b).

Thus it is the ability of certain groups within the City of London – accountants, lawyers, bankers and screen traders – to superimpose their rhythms on others which gives the impression that they are the only ones present. The construction of space in their own likeness through a series of rituals, gestures and mannerisms serves to empty out the spaces of accountancy, law and money of any economic activity other than their own. Codes, as Lefebvre points out, 'function durably, more or less tacitly and ritually; they organise and rhythm time as well as relations' (1996: 234). Through a constant succession of movements and activities, the manner and style in which they are executed, as well as the mood and tempo they generate, the above groups are able to dominate space in their own image.

Excluded from the 'membership' of such a space, then, are those whose rhythms and movements do not accord with the dominant representation and use of such spaces; namely, the routine service-workers who pass unnoticed through the offices and buildings, and the ancillary services who keep the physical operations up and running and the infrastructure intact. But as I have had cause to emphasize, such groups are not physically excluded; rather their presence is *smothered* by a dominant coding of space which takes its cue from finance. At its most obvious, amidst the animated gestures of those buying and selling, the rise and fall in the volume of conversations and voices, the routine lulls between frenetic activity as markets dip and rise, and the banks of screens strewn across the trading rooms, the daily routine of anyone other than screen traders remains largely unseen and rarely acknowledged. Even when trading is over for the day and the floors are silent, it is difficult to conjure an image of such places as about anything other than finance.

Such is the power of coding in these financial spaces, therefore, that, in Lefebvre's terms, users other than those involved with finance directly are filtered out through the lens of homogeneity. More to the point, the image of homogeneity created around banking and finance is something that is accomplished through the spatial practices of the traders and their associates, rather than simply imposed by the so-called institutional power of finance capital. The exclusive nature of the City as a place of finance which generates its own sense of 'membership' through the sound, sight and even scent of those recognized as present is thus, in itself, an achievement. It is the outcome of a series of temporally and spatially discrete practices which, as Lefebvre would have it, amount to the exercise of a dominant presence.

Alongside, or more accurately, *within* this domination of space, however, Lefebvre, as we have seen, also wished to argue that those 'out of place', those who pass unrecognized, are none the less able to inhabit it. Or rather, they are able to appropriate spaces for themselves within the dominant coding and use of space. On this view, the most effective appropriation of spaces is made by those who make symbolic use of what is around them and turn it to their own needs, either by subverting the codes of the dominant space or by representing an alternative way of inhabiting it. Such spaces are first and foremost 'lived spaces', in the sense that it is the routine practices and the ways in which space is related to, laid claim to or personalized even, which may undermine or challenge the dominant representation of space. It is in this more expressive vein that they may contradict the formal representations of space through a spatial code which is neither simply read nor imposed, but used as a means of living within that space (for an elaboration of this point, see Lefebvre 1991a: 362–5).

So within the old and new buildings of the City, for example, among the formal spaces of trading and financial dealing, there are spaces in the interstices occupied by routine service-workers which have no formal presence or recognition. These may be fixed or semi-fixed spaces, awkwardly shaped or lost in the bowels of buildings, where a newspaper can be read or the radio listened to, or they may be passageways, walkways, agreed places to meet which offer some respite and an alternative to the formal coding of banking and finance. Messengers or tradespeople, cleaners or security guards may use such transient spaces to meet their own needs, to lay claim to some attachment to their surroundings, as a means of deflecting the singularity of the dominant coding.

None of this is meant to imply, however, that the different forms of representation simply coexist, with one form of social space merely subordinate to the other. According to Lefebvre, the power of abstract space to erase the traces of others, to reduce difference, is never entirely effective. The very attempts to achieve homogeneity do themselves produce spaces which 'escape the system's rule' (1991a: 382). Thus, for example, it is possible to see the actions of those who move through and across the daily activities of finance as themselves capable of subverting a dominant coding simply by going about their own business in their own particular way. The business of waiting at table in the director's private dining room at one of London's top banks may serve as an illustration.

Waiting at table in the private rooms set aside for the sole use of directors is a useful example because much of the work involved is performed alongside or in front of the directors and their formal use of the same place. Yet the manner in which waiters inhabit the same place may escape the order imposed by the formal routines of the directors. Submissive in their conduct, the waiters and butlers may none the less work more to their own professional code of presentation than to any expectations of the directors and the institution at large. This may, for example, concern the general presentation of the table – the arrangement of the cutlery and napkins, as well as the polished shine on the glasses – where each detail is suggestive to other waiters but not necessarily to those seated at the table. In this instance, the lack of focused co-presence between diners and those who wait on them, which speaks to a formal agenda of servility, may actually be used to meet a quite different set of professional needs to those of merely finance. Although the waiters lack the formal means to negotiate the dominant coding, they are able to deflect it through their own spatial practice. The two worlds of business are entangled, yet remain far apart.

In Lefebvre's terms, such spatial practices do not amount to a wholesale rejection of all that is signified by a particular dominant space. Rather, such spaces remain internal to the dominant form ('induced' rather than 'produced' space), brought about in the above case by the requirements of a specific division of labour. Whilst such routine service-workers remain part of someone else's space, the manner in which they inhabit that place nevertheless leaves them free to imagine themselves differently from those involved in the daily grind of finance and its marks of distinction and, indeed, to use the space differently.

But if this is so, what does it mean exactly to talk about the *domination* of space in this instance, or in any similar situation where social groups routinely occupy space in such a contradictory, often uneasy manner? Within the worlds of work, be they manufacturing, service or otherwise, it is not hard to envisage similar contradictions of space, to use Lefebvre's term, where there is a clash over the meaning *of* space. In fact, many a political tension over the use and social meaning of public space would fall under the rubric of spatial contradiction where the attachment of one group clashes with the formal designations of another.

In such cases, where the prohibitive aspects of space are exercised in all kinds of ways and the room for manoeuvre and 'escape' from them remains a real and lived possibility, why should all this social activity be reduced to one thing, the domination of space? Domination, as has been stressed throughout these pages, from Weber's account early on, is about the closing down of possibilities, where those who are formally free to do otherwise have no choice but to fall into line. The removal of choice and the imposition of constraint lie at the heart of the process, to the extent that it is made difficult for groups to do anything other than submit to the dominant will. Such parties may be free to use space in all manner of ways, but 'objective circumstances' dictate, if that is the right word, that they submit to the dominant coding.

To be blunt, though, the interplay of social forces that Lefebvre claims as lying behind the tensions and contradictions of social space is hardly captured by the scenario of domination and resistance. On the face of it, rather more seems to be going on in the name of power.

What in the name of power?

In his exposition of Lefebvre's account of modernity and space in *Geographical Imaginations* (1994), Gregory seems to understand that there is rather more involved than simply the sheer domination of space. Recognizing the range of influences upon Lefebvre's spatial thinking, and one in particular – the radical critique of everyday life developed by the Situationists in France and Guy Debord in particular – Gregory wishes to convey the experience of what it means to have one's spatial presence smothered or erased by an 'occupying' force. The terminology is deliberate, for it conveys for him the sense in which Lefebvre thinks about everyday life as a 'colonized' experience: as an occupied set of spaces where attachments are broken and meanings severed.

Central to this is the bureaucratization of space, where the spaces of the everyday are mapped out by administrative systems and subject to routine surveillance and regulation by the state. If this is somewhat broader than the worlds of service work described, it none the less conveys the sense of dispossession and displacement felt by those whose daily life is routinely confined and constrained by an 'occupying' force whose signs and signposts map out a rather different 'territory' from their own.

Within this colonization of everyday life, however, Gregory recognizes the importance that Lebfebvre attaches to the imaginative and often spontaneous challenges to the more formal codings of social space which inevitably occur. Drawing upon the legacy of Surrealist thought and play, the possibility of the subversion, disruption and reappropriation of everyday spaces remains a constant threat and ever present tension to the superimposed order.[3] The room for manoeuvre, the gaps in the formal coding and the ability to lay claim to some reattachment of meaning, no matter how tenuously perceived, are a firm indication of the kind of surface collisions involved. However, they are also an indication that recourse to analogies of colonial domination to describe the broad indeterminacies of power-laden social space may be stretching the point too far.

---

3 Gregory draws attention to the distinction between 'spectacles' as formal, staged events and the more expressive, immediate idea of 'festivals' that Lefebvre often returned to as a way of conveying an authentic moment which challenged the 'colonization' of everyday life (Gregory 1994: 405). There is an idea of meaningful play here, which can have a disruptive effect simply by tapping into the embodied and imaginative side to festival-style activities. Lefebvre's critique of staged, spectacular forms of action was influenced by Guy Debord's work, which attempted to harness spontaneity and passion to the side of political (re)action. See Shields (1999) for an account of the often tumultuous relationship between Lefebvre and the Situationists up to May 1968 in France.

For simply to speak about the messy coexistences and juxtapositions of power which characterize much social space in terms of colonization, replete as the word is with references to occupation and reterritorialization, is still to adopt a rather broad brush treatment of power. While the term usefully draws attention to the genealogy of Lefebvre's thinking, in Gregory's hands it shifts us more towards an enframed, ordered understanding of social space than is appropriate given the more open arrangements of power involved.[4] If we look again at the prohibitive aspects of social space referred to in the City of London illustration and their relational character, neither domination nor its spatial proxy colonization captures the provisionality and indeterminacy of the arrangements involved. And that, quite simply, is because there are other modes of power in play producing a prescribed sense of social space. In different arrangements, there are different modal alignments of power, and in the jumble that is the City as a place of finance, authority as well as domination, seduction as well as manipulation, each has in its own way put people in place.

First up, it is possible to see how the exercise of *authority* has shaped what it means to be in London's Square Mile, both for those actively engaged in the work of finance and for those tangential to it. In referring earlier to the City as a site of cultural authority, one could be forgiven for thinking that it just is; that its global status as a financial authority is a given. Yet that would be to gloss over the fact that such an authoritative space is actually *produced*.

This production of an authoritative space can be thought about in two related but distinct ways; one in terms of the City as a recognized centre for financial expertise and the other in terms of it as embodying a recognized style of authority which places finance at the apex of economic activity in the Square Mile. Whereas the former may be considered as a claim to expertise based upon the repository of all past and present financial knowledge within the confines of the City, the latter may be considered as a exercise in securing assent by those *in* authority. If the position of the former requires that their competence is acknowledged before authority is ceded to them, the position of the latter is dependent upon others deferring to the greater importance of their economic role. Neither position is fixed, however; rather the authority 'held' is performed in a variety of ways and settings which mobilize the surroundings as a prime resource.

So, if we were to push Lefebvre a little further, it would be possible to see that a number of the coded spatial practices referred to earlier effectively

---

4 The notion of enframing developed by Timothy Mitchell in *Colonizing Egypt* (1988) seems closer to the idea of colonization that Gregory has in mind, where social space is ordered in a legible, picture-like way that makes it 'readable' to the dominant presence, and thus manageable as an administrative, controlled unit. As an account of power relations behind the production of everyday social space, however, the colonial analogy, whilst evocative, is far too reductive in Mitchell's Foucaldian-inspired version.

establish the City as a space of authority, in many of the ways discussed in chapter 5 in relation to Richard Sennett's work. Those who act as *an* authority in the City draw upon wider networks of expertise to bolster their credibility in the eyes of others and, in so doing, draw upon the trappings of past and present resources to confer legitimacy. Ritualized ways of doing business conducted in both a traditional and a modern manner lend a certain weight to the views expressed and give a voice authority. Weight is also achieved in other ways, through the monumental buildings of the Square Mile (the Bank of England or the Mansion House in the City, for example), which serve as a symbol of authoritative values and judgement, gained largely through the respect that comes with the passing of time. Authority in this context is thus primarily established visually: through the splendour of the buildings and their facades, their obvious intent to impress and, equally, their inaccessibility. Once inside, the weight of the past may be measured just as easily by the ear, by the play of sounds, floating voices and muffled silences.[5]

Alongside this demonstration of (an) authority it is also possible to locate the coded practices of those *in* authority. Less concerned with the symbolic trappings of authority, those in authority owe their position to the organizational structure of which they are a part. By virtue of their corporate organizational role, as bankers, brokers, traders, dealers or whatever, those involved in the day-to-day business of finance assume an authoritative presence that is dependent upon others recognizing their position and, indeed, social space. The coded styles of authority, the gestures and mannerisms which convey to others that they have control over particular zones of economic activity, on this account, enable them to superimpose their daily rhythms on City space. It is the proximate styles of control involved and the embodied use of social space which effectively add up to the production of an authoritative space.

At this point, though, perhaps we need to ask again whether that is all that is going on in the name of power. If it were simply the case that the City of London's prohibitive spaces could be accounted for by relations of authority rather then domination, then the assent secured from above by those in authority or the acknowledged weight of expertise practised by the guiding authorities would end the story. Yet, as Lefebvre's stress upon the

---

5   Jonathan Rée's *I See a Voice* (2000) draws attention to the spatial indeterminacy of sounds in comparison to visual effects, and the less materialistic nature of the former, which gives it a certain ethereal quality in monumental buildings. 'Hearing does not presume as much as vision. It is not so arrogant, and it is willing to refer its experiences to evanescent qualities without insisting, as sight does, that they have to be tethered unambiguously to definite things in the material world. The ear is not hung up on the reality of things: it is happy to live modestly in a world of sounds which, as Proust wrote, "have no position in space"' (2000: 46).

contradictory nature of social space forewarns us, recognition and compliance are not wholly the issue. The embodied clashes over the inscribed meaning of space and the tensions evident between those who inhabit the City as a workplace point to the presence of other sets of power relations; the most obvious of which is manipulation.

Manipulation, as noted before, is all about concealment and indirectness. If certain groups within the Square Mile are able to construct a series of spaces in their own likeness by filtering out the traces of others, then that may just be because indirectly they have inculcated codes of conduct which appear inclusive. The fact that such codes may conceal their motives or disguise what, in reality, is a place of global business and commerce, and only narrowly that of finance, is at bottom an act of manipulation. Neither compliance nor competence is the issue when spaces are coded in such a way as to mislead people into imagining that they are all part of one cultural space, when in fact their own particular attachments have been selectively erased. In such circumstances, there is no real possibility of challenging or resisting the manipulation of codes, principally because those on the receiving end would be quite unaware of the deception involved. Of course, if the act of manipulation were to be revealed, then the embodied clash of meaning would be all the greater precisely because of the deception. In consequence, any remaining vestiges of authority would also very probably disappear rather abruptly once recognition had been withdrawn.

Suggestively, at times Lefebvre seems to come close to the idea that it is the manipulation of spatial codes which is at issue, in particular when he talks about social and spatial differences routinely suppressed by those who attempt to speak for the whole, yet interestingly the end result is always that of the domination, not the manipulation, of space. It is as if the mischief done by thinking that domination is everywhere stops him from taking the idea of concealment seriously. Equally, what for certain does not appear in Lefebvre's vocabulary of power, and perhaps all the more surprising given his noted distaste for formalized spectacles, is *seduction*.

In thinking about how certain groups are able to leave their imprint on the City as a place of finance and indeed *represent* it as such, the seductive nature of many of their actions is not particularly difficult to discern. Among the more obvious acts is the ability to represent the speed and magnitude of monetary flows as something exciting and attractive to be around. Who, after all, does not associate the buzz of money and the City with the mood of the present? Less boxed in by the ponderous weight of authority and the trappings of the past, the images of finance projected testify to the mobility, risk and individual freedoms associated with a 'fast' lifestyle of dealing and speculation. In comparison, other forms of economic activity are made to look slow and leaden. The very power of the jazzy representations lies with their suggestiveness.

Undoubtedly, not everyone is taken in by such images, but it is precisely the possibility of rejection or indifference as a response which gives seduction its characteristic reach. Seduction, you will recall, is a form of surface power which acts upon the choices of those who have the ability to decide otherwise. In taking advantage of attitudes that are already present, it capitalizes upon those tendencies which already appeal and seeks to proliferate them. 'Membership', in this instance, is all about imposing a pace and a rhythm on to the City's economic life that is suggestive of a fast tempo and a mood of spontaneity. Being made to look slow is to be represented as 'out of place'. The preoccupation with speed that takes its cue from the frenetic activity of the marketplace and the animation that mirrors the rhythm of those buying and selling both work to exclude, although on this view less by the force of domination and rather more by the force of suggestion and presumption.

Let me be clear about the nature of my concerns here. In criticizing Lefebvre's inability to see anything other than the domination of space, and Gregory's related endorsement of that position though the metaphor of 'colonial' occupation, my point is simply that there is so much more going on in terms of *what* power is exercised across the colliding spaces of the City. It is not, I should stress, that I wish to say that seduction rather than domination is a more accurate portrayal of what happens on a day-to-day basis in City life, or that manipulation not authority is what keeps people 'in their place' in such surroundings. Rather, my argument is that in places like the City of London, power is likely to be exercised in a variety of ways, from domination through manipulation to seduction and even down to coercion. This is not because power is endowed with some kind of vapid plurality, where the different modalities take it in turn to act out their display. On the contrary, it is the nature of the places themselves, how they are *constituted* through the practices and the rhythms of the different groups which inhabit them, which gives rise to tangled arrangements of power and their execution.

In the case of the City of London, for instance, the experience of power for many of those who work on the edges of finance, as support staff or ancillary workers, is likely to take in both acts of domination and manipulation, as constraint operates alongside concealment in the routine suppression of their visibility. Such acts are merely different ways of achieving the kind of smothered presence that Lefebvre brings to mind. Equally, the monumental buildings of the Square Mile, such as the fortress-style Bank of England, do act as a source of authority, yet given the changing role and function of the latter institution, it is probably now more accurate to talk about its persuasive role in the contemporary world of UK finance. The meaning and representation of institutions, their rituals, practices and architectural styles, change, reflecting the altered make-up and composition of those who use them. It is *not* the case, therefore, that such institutions represent pre-formed blocs of power into which various groups are then slotted. Rather, the mixture of groups which inhabit various institutions, and the relationships which establish their *presence*, imbue such institutional spaces with power. And it is from the specific relational ties, the interplay of forces that lie behind the comings and goings of those present, that power takes it diverse modal expressions.

Closed Worlds and Open Walls

Although Lefebvre remained largely oblivious to the nuances of power and social space, he was hardly alone in equating domination with power and seeing its traces all around him. Perhaps the more familiar sense in which domination is considered as pervasive stems from the notion of closed spaces; spaces constructed by groups building 'walls', sometimes literally, to exclude those who are not 'the same'. If, up to this point, we have been thinking about the cross-cutting nature of people's lives as they go about their daily business, the idea of spatial enclosure suggests a fixed separation, a boundary line in the sand so to speak, which produces a clear limit to the movement of 'others'. In place of messy coexistences and jumbled lives, closed spaces are about groups 'walling themselves in', erecting social and physical barriers to the comings and goings of others.

The obvious example that springs to mind is the kind of spatial enclosure known as 'gated communities', which, since the 1970s, have effectively partitioned sections of many of the world's major cities. Typical of many North American cities today, gated communities are more or less controlled enclaves, private spaces which people buy into for their own exclusive use. The claim to exclusion is based upon living behind, say, a three-metre-high wall where all the residences focus inwards so that the outside world is 'filtered out' of the imagination. Cost is obviously a critical factor in deciding who lives in such closed communities, but on an everyday basis control over who enters and who mixes with whom is exercised through surveillance cameras, uniformed guards and the like. Such closed spaces are highly ordered, offering their residents a rather different sense of 'membership' from that outlined by Lefebvre.

If, for Lefebvre, a sense of membership is something that is produced by a group constructing a social space in its own likeness, in the gated communities described by Teresa Caldeira (1996) any likeness is formally imposed by strict rules of conduct. A sense of who belongs is achieved not by a collective construction of who is recognized as present, but rather by a set of rules which imposes a like-mindedness. Membership of such communities is delimited sharply so that anyone wishing to be part of it must conform to the prescribed rules of conduct – which can include anything from the external decoration of dwellings to the layout of gardens and to regulations for 'noisy' or other forms of 'inappropriate' behaviour. Such rules are intended to dramatize the difference between those on the 'inside' and those on the other side of the walls. Here difference is not so much smothered as physically barred, and sameness formally imposed rather than socially constructed.

It is perfectly accurate, therefore, to describe such closed spaces as spaces of domination, not merely because they exercise strict control over entry on the grounds of cost, but also because the formal rules constrain the behaviour of *all* concerned – those who live in the communities as much as those who pass through them, as servants, security guards, gardeners and the like. The closing down of possibilities, the restrictions that residents have to abide by in order to be part of 'the community', leave them with little choice but to submit to the formal order of things. Residents may lay claim to spatial exclusivity over others, yet their bounded 'defensible space' also serves to limit their own freedom.

Cultural separation, the act of self-enclosure, is not only a characteristic of rich communities, however. As a defensive reaction, poor communities may 'wall themselves in' as protection against the more economically or politically powerful (in terms of resources). Ethnic groups may opt to draw a line around their community to avoid discrimination or to preserve an identity perceived to be under threat from the 'outside'. Such acts are rooted in Hannah Arendt's mutual action, the solidaristic networks that rest upon the mobilization of collective resources. Rather than think in terms of power as instrumental leverage, such acts of closure take their cue from the 'power to' protect themselves, not from some desire to exercise 'power over' others. As a result, any notion of order is more likely to stem from informal and negotiated practices rather than prescribed codes of conduct.

Interestingly, the attention paid to the regularized, almost disciplined, styles of behaviour in the rich, gated communities is somewhat reminiscent of Foucault's early work on diagrams of power, discussed in chapter 4. The attention paid to the zoning, parcelling and enclosure of space as a means of regulating, as well as enabling, movement through a variety of institutional complexes, from hospitals and schools to military barracks and prisons, sounds analogous to the 'techniques' of domination described by Caldeira. The analogy breaks down, however, principally around the fact that for Foucault the techniques of power show up only as an effect on the actions of others. There are fewer direct or overt restrictions on people's behaviour and rather more actions which seek to *induce* the appropriate conduct in others. Here, we are already moving away from a strict notion of enclosure as the domination of space to a consideration of other possible acts of power; in this case inducement, but in others the possibility of indirect manipulation or even arm's-length seduction.

From my argument so far, you might well ask how it could it be otherwise. To suppose that even closed spaces have the last word on domination would be to gloss over the different ways in which closure can be achieved. But my point here is a rather different one from simply drawing attention to the fact that power is exercised in subtle and not so subtle ways. It is that much of what we take to be closed space is usually less closed than it seems, and much of what appears open and accessible is not always so. If Lefebvre's account of the production of social space has persuaded you of the latter, then a more dynamic sense of space which foregrounds both movement and circulation may convince you of the former.

Most bounded spaces, it seems to me, are relative rather than total and whilst some borders have hardened over time (notably around migration controls in Europe and the US), many of the everyday spaces which make up people's lives – from shopping malls, office spaces and housing estates to public parks, airport lounges and a whole slew of public institutions – are less obviously impermeable. Even in closed, gated communities, boundaries are routinely crossed by all manner of tradespeople and public officials, as well as friends and relatives, and the supposed stark lines of difference are compromised daily by the service rhythms of domestic care, maintenance and security staff. 'Open walls' rather than 'enclosed worlds' is perhaps a more apt metaphorical redescription of the boundaries of such places. The porosity attached to places is something that Doreen Massey (1994), in particular, has long been keen to stress, and what we can take from this observation is that, in many a familiar place, people move in and around one another, sliding across each other's lives, and establish a presence through interaction in all kinds of powerful and not so powerful ways.[6] And such interactions, I would argue, disrupt any easy cultural mapping of who is close at hand and who is distant, who belongs and who does not.

Open walls

Lefebvre, as we have seen, cast this kind of interaction in the mould of a domination/resistance relationship, which undoubtedly is the case in some

---

6 I take this sense of power and place to be in line with Doreen Massey's mobile power-geometries of space-time, where the very articulation of power relations involves the movement through and use of space. See Massey (1993, 1994, 1999). The idea that spatiality is imbued with power is common to both understandings, although the stress that I place upon the presence of power relations moves the analysis closer to a phenomenological approach where routine practices, rhythms, meanings and attachments occupy the foreground. The other, more obvious, difference between the two accounts of power and spatiality is the importance I attach to the distinctive modalities of power and how they are spatially circumscribed.

situations where the interplay of forces shapes that particular relationship. In the private spaces of the City of London's financial institutions, it is quite conceivable that domination played its part in erasing the traces of those only tangentially connected with finance and banking; although, as I have argued, there was more going on there in the name of power than simply domination (or resistance for that matter). So, in a more open, forgiving, public space where people mill around and cross one another's paths in largely unforeseen ways, we could be excused for thinking that power is largely absent from the tangle of associations. That, however, would be to conflate accessibility with openness.

Many an open space may be closed down by degree, and by that I am not referring to the all too obvious point that access or entry may be denied through the imposition of formal restrictions. Rather power may be exercised in unbounded spaces in far less marked, indirect or even shallow ways to achieve the desired form of closure.

At the shallow end of power, for instance, the use of electronic surveillance technologies to monitor and track the movement of people in public places is perhaps the best-known way in which our conduct may be subject to scrutiny and possible reproach. But much depends upon to what end such technological mediums are supposed to meet. In themselves, technologies like closed-circuit television cameras (CCTV) are little more than a resource which may generate an effect we experience as power. They are, as was argued in chapter 5, merely part of a range of mediums through which power may be exercised. They are not powerful in themselves. In general, the impact of such surveillance technologies has been overblown, with far less attention paid to what is actually done with the visual data amassed than to the fact of installation itself. In many ways, the intended effect of CCTV technologies, as Foucault would say, is to induce people to regulate their own behaviour in line with what is expected of them in relatively open places. That may well be so, but in the absence of knowing what use is made of such technologies, their impact remains indeterminate. They are the means of power rather than power itself. In other words, they are power in name only.

A more intriguing and indeed less distinct example of the exercise of power in open settings is where suggestion, not surveillance, holds sway. This, you may recall, is one of the hallmarks of *seduction*, where curiosity rather than any so-called disciplinary logic is the subject of stimulation. Earlier in the chapter, I spoke about seduction in the context of the enticing ways in which the work of finance in the City of London is represented. In seeking to take advantage of attitudes already present, the inviting glamour of financial games-playing acts as a lure to those predisposed to such a message. The fact that the world of finance, and the world of the City in particular, may convey such a message only in partial ways is neither here nor there. Seduction is about the work done on such possibilities, imagined or otherwise.

Similarly, in open spaces where there are no formal restrictions on access, such as those which mix shopping and browsing with relaxation and entertainment, seduction may be at work, offering possibilities and creating the illusion of access. Closure in this context is about distraction, both in the sense that our desires and interests are indulged in selective ways and in the sense that we remain unaware of the scripted nature of our movements. Perhaps an example will allow me to elaborate on the actions involved.

In present-day Berlin, much has been made of the fact the city is once again at the sharp end of modernity, intoxicated with the new, unsure of how to negotiate the past, and waiting to see what the future holds in store.[7] In many ways, the new national capital is overdressed with symbolic meaning, from the monumental glass and brick-clad towers at Potsdamer Platz and the elite fashion spaces along the length of Friedrich Strasse to Norman Foster's glass-domed gesture to democracy, the Reichstag, as well as the many statements of a Prussian past echoed in the architectural make-over of prominent buildings. Of these, it is the monumental form of the buildings and plazas at Potsdamer Platz that I wish to focus upon, as they offer a series of interesting contrasts and overlays in the way that their quasi-public spaces articulate power.

Potsdamer Platz is dominated by two developments, the Debis quarter and the Sony Centre, both of which offer a mix of high-rise office buildings, shopping and leisure facilities. The former development is a testament to the global financial ambitions of the Daimler Chrysler corporation and the latter, considered here, more obviously reflects those of the Sony corporation and its pre-eminent position in the global entertainment industry. Of the two complexes, Sony's range of consumer offerings – bars, restaurants, a style store, an urban entertainment centre complete with IMAX 3D cinema and an eight-screen cinema bloc – leaves little doubt as to who is its corporate owner. The place itself acts as a kind of exhibition complex for the entertainment wares of Sony PLC – from Sony PlayStations and online movies to all manner of electronic wizardry – effectively 'branding' the space as a vehicle for consuming its archive of films, music and entertainment software.

---

7   Alexandra Richie's *Faust's Metropolis: A History of Berlin* (1999) is a vivid account of how Berlin has borne the marks of nineteenth- and twentieth-century history and found itself successively renegotiating its recent past. For a sense of how contemporary Berlin has undergone a symbolic re-enactment of place, see especially Nicholas Howe's (1998) account of the city as 'history interrupted by construction' and Andreas Huyssen (1997) on 'the voids of Berlin'. See also Allan Cochrane and Andrew Jonas (1999) on reimagining Berlin.

At one level, the layout reflects the decidedly rational organization of the cultural experience on offer; its meticulous, uniform nature with its roots firmly in the commercial logic of capitalist calculation. Yet on another, more outward, level, the space itself is given over to pure indulgence, fragmented sense impressions and dream-like qualities. In this self-styled space of 'edutainment' the experiences on offer represent a contemporary form of escapism which articulates a modern sense of distraction; one that for some may be empty of meaning, yet is no less fulfilling because of that.[8] Its register is not so much the wholesale recreation of entertainment values and attitudes as pleasure, relaxation and, of course, seduction. To move through the complex is to find oneself subject to a power whose imprint is decidedly modest, where spontaneity and impulsiveness rather than any systematic stress are the pulling force. At best, the experience generates an interest in Sony's merchandise, perhaps reinforcing a preference for its brand of goods over its competitors.

Part of this pull, this stimulation of curiosity, stems from the design of Sony's elliptical central plaza, a generous space laid out under a dramatic, tent-like roof structure. Constructed out of glass and fabric, and illuminated at night, the monumental form of the roof draws the eye down into the composed open space below. In contrast to the rather heavy-handed commercial spaces of the Debis complex, the quasi-public plaza is an invitation to mingle, circulate, loiter and perhaps browse a little before searching out any entertainment that is on offer. In this way, the navigable space works through an atmosphere of detachment, much like any urban street, yet at the same time it provides a glimpse of what else may be absorbed and consumed. This, then, is not the visual establishment of authority through the splendour of a monumental building, as is the case in the City of London, but an act of seduction where the suggestiveness lies in the layout and design of the social spaces themselves.

As always with seduction, the invitation can be declined or ignored; people are free to walk through the place, to take a cup of coffee in its cool surroundings and leave. Openness is equated with access and the

---

8 The phenomenon of distraction itself was subjected to scrutiny by Siegfried Kracauer in his writings on the cultural spaces of Weimar Berlin in the 1920s and 1930s. In some of his better-known essays (collected in Kracauer 1995), 'Those who wait', 'The mass ornament', 'Cult of distraction' and 'The little shop girls go to the cinema', he reworks the theme of cultural lack and loss of meaning to present a more ambivalent attitude to the spectacles of modern mass culture. With pointed reference to the products of 'American distraction factories' (such as the Tiller Girls, a highly drilled dance unit analogous in Kracauer's mind to the formless abstraction of Taylorist production) in particular – mass cinema, theatre and dance – he attempted to show how this realm of distraction had become a *necessary* reference point for an understanding of the modern condition, not some whimsical tangent.

choice to opt out of the 'show'. Or so it would seem. The choice to walk away is there, of course, but the notion of unconditional access is partly illusory.

Seduction is an instrumental mode of power primed to shape and mould the will of the many whilst allowing individuals the possibility of opting out. It is about suggesting this rather than that option, and turning an apparently open-ended situation to particular advantage. In this respect, the Sony plaza at the heart of the Berlin development may well be open to the public but it remains a privatized space, closed at midnight, and laid out in such a way as to script the movements of those who walk through it. The movement and circulation of those who freely choose to enter is focused, despite the impression of abundant space suggested by the inner plaza. The 'paths' through the complex are less formalized representations of space, in Lefebvre's sense of the term, than invitations to go this rather than that way, to follow the glimpses of entertainment rather than simply walk away. A seductive presence is apparent from the combination of suggestive practices, experiences and spaces that are laid out for temptation. And it is through such largely unmarked presences that an open complex like Sony's development may indeed be closed down by degree; although in this case by a process of inclusion rather than exclusion.

In line with the phenomenological approach to space set out earlier, the varied spacing and timing of people's interactions implies that for power to have a presence it does not always have to take the form of a physical or social barrier. Power is not only characterized by claims to spatial exclusivity. As before, the different arrangements of power possible take their shape from the placement of forces and their relational ties. Some arrangements, where a claim to domination rests upon enclosure behind a tall fence, are rather obvious to discern, whereas in other, more open spaces a less obvious arrangement may involve acts of seduction run alongside, say, inducement or even manipulation, where access is concealed rather than denied. Quite simply, the mutability of power differs in line with the differences between places, in terms of their uses, attachments, codes and relationships.

But, and this is an important but, in so far as each and every relationship is not a relationship of power, so each and every place is not continuously marked by the presence of power.

In the Presence of Power

The idea that power turns up more or less everywhere is a familiar one, now that Foucault's immanent conception of power has gained something of a foothold across the social sciences. At its most general, the idea has much to commend it if it succeeds in undermining the steadfast notion that power comes in blocs and is always centralized. Unfortunately, familiarity can also dampen the very critical faculties that made such an idea so useful in the first place. To say that we are all more or less immersed in arrangements of power does not mean that it is ever present or that the durable architecture of power is all around us waiting to put us in our place. True, power has to have a presence if it is to be at all effective, but that presence is not a ubiquitous one. There is no everywhere to power. Not all places are saturated with the trappings of power, and power has not been scattered and dispersed to such an extent that few, if any, are outside of its reach and scope.

It comes back to the *particularity* of power and to the discussion in chapter 5 which stressed the need to pin down power relations in their various guises. Power is always exercised in particular ways, through various modalities, and establishes itself through specific relational ties. If it has a presence at all, then that is because it makes itself felt through the interplay of forces *in place*. As was said before, social groups mark their presence in particular places through a variety of relations and practices. Whether or not such relations are relationships of power is an empirical question and stems from an understanding of the substantive qualities of power: that some relationships do close down possibilities, conceal choices, impute threats, reward compliance, and some relationships simply do none of those instrumental things.

Thus far, in this chapter, we have considered the notion of a powerful presence in a variety of ways. First, by drawing upon Lefebvre's understanding of social space it was possible to explore how a *smothered presence* in the institutional spaces of London's Square Mile was achieved, in this case largely through acts of domination and authority working in tandem. The effective construction, or rather I should say successful representation, of a blanket financial space by bankers and the like in their own image arose precisely because of the specific placement of forces. If such forces were to alter or be radically reconfigured, then so too would the sense of presence, from a dominant presence to one more in tune with the new play of forces.

A second, more demonstrative form of presence was explored through the example of the physical or social barriers erected by residential communities to enclose their social space. Here a *barred presence* was established principally through the practice of domination, although it is entirely possible that such acts may be underlined by the threat of coercion or outright sanctions. Again, it depends on the forces in place and the nature of the differences made between those on the inside and those on the other side of the 'wall'. And in the third, rather contrasting, example of an unbound, relatively open space, a more modest *unmarked presence* was explored, in this instance Sony's quasi-public space at the heart of Berlin's new development in Potsdamer Platz. Here, through the suggestive pull of the layout and design of the new complex, it was possible to see how a seductive presence effectively closed down options, enticing visitors to circulate in ways that they might otherwise not have chosen to. In each of these instances, therefore, it was possible to show how the establishment of a different type of powerful presence arose from the forces *in place*.

Left at that, however, there is nothing especially compelling about the presence of power. If the point is merely to register that such forces are present in the here and now, that to all intents and purposes they are here rather than there, then the analysis amounts to little more than a plotting exercise. But there is more to being in the presence of power than simply that.

More or less present

We shall see in a moment the significance that past states and associations can have for the establishment of a powerful presence, but first I want to assert the rather unexceptional point that presence is about more than simply 'being there'. In the previous chapter, I spoke about the varied relationships between proximity and presence, and the possibility of institutions making their presence felt at a distance. In particular, I questioned the extent to which it was possible for today's big financial institutions to exercise a simultaneous, controlling presence over globally dispersed money markets. My argument, you may recall, was not that power at a distance is implausible, but that the kinds of institutional arrangement exercised by the large finance houses invariably mixed direct and indirect styles of power, far-reaching with more proximate modes, to stand any chance of achieving an influential presence world-wide. Domination at a distance, courtesy of the new-found information and communications technologies to achieve near instantaneous reach, is something of a caricature; more an exercise in impression management than the real thing.

What there are, however, in such situations are relations of co-presence, where it is possible to be on the other side of the globe and still have a part to play in local affairs. The point, quite simply, is that you do not have to be present to have a presence. In the example of financial markets, it is possible to have a powerful presence yet remain far off, at a distance from where any number of significant events may be playing themselves out in real time. Presence, in this sense, is merely the flip-side of reach; another way of conceiving how it is possible to traverse or negate the gap between here and there.

At a more general level, however, what we can draw from this example is that the exercise of power in particular places may well originate beyond those places, at some other location, yet remains part of power's active presence. In other words, the power relations in place are affected by what happens elsewhere and the network of connections of which it is a part. Thus in thinking about places as a product of the messy coexistences and interactions which shape them, any placement of power involves the distanciated actions of those physically absent as much as it does those present and in close physical proximity. What puts people and power in place, however, is the mixture of styles and modes peculiar to a particular location.

So, in considering once more the City of London as a centre of 'cultural authority' in the way that its financial analysts draw upon wider networks of expertise to gather, circulate and interpret information, their authoritative presence is established *both* through their global business connections *and* through the everyday codes, gestures and mannerisms which they use to bolster the weight of their claims. Such claims to authority, as we have seen, may be fused with a more manipulative presence, where those not directly involved in global finance may be misled into believing that they too are part of the *one* global financial space. Or such claims to global authority may stand alone. It depends. That is, it depends upon what basis a powerful presence is established and, indeed, maintained, not all the elements of which will be present in both time and place.

One such element is history itself, or rather the way in which the past inserts itself as an active trace in power's presence. What is selectively called up and mobilized from the past may readily act as a resource through which power is exercised.[9] The notion of resource here, however, is not something akin to a linear history which can be sliced up and drawn upon to inform the present. Rather, the past as a resource has an ongoing presence where associations and memories may figure in what, for example, it takes to establish an authoritative presence, or indeed any powerful presence for that matter. This may take a straightforward form, as for instance in the way that 'tradition' has selectively informed so many of the gestures and mannerisms that have helped to keep in place the style of authority practised by long-standing 'members' of the City's financial community. Or the presence of the past may irrupt into the present in less invented or predictable ways. The monumental forms of the Potsdamer Platz in Berlin are just such a case.

---

9 Mike Crang (2001; Crang and Travlou 2001) draws upon Deleuze to make the point that Being should not be confused with being-present. Arguing for a fuller sense of place, one which leads to temporalizing places rather than spatializing time, Crang provides a sense in which it is possible to think of times as mixed, looped and folded into places. On this view, the past is not something already there and waiting to be reflected upon and incorporated into the present; rather it takes the form of a series of associations and possibilities which may be brought to bear on a place in the present.

One of the striking observations about the Sony and Debis developments at Potsdamer Platz is the extent to which Berlin's past has been air-brushed out. There is no indulgence either in the symbolic significance of the site itself, as an historic interchange at the heart of Berlin's city life before the Second World War, or in a style of architecture reminiscent of Berlin's Prussian past. The conscious decision not to celebrate a Prussian style, utilizing local stone and low-line development, in favour of a clearly global brand of glass, aluminium and steel walls is a statement hard to miss. But that does not mean to say that the past is absent from that statement. On the contrary, by consciously writing out known, selected moments of Berlin's recent past – from Bismark's imperial Prussian past, to its Weimar and Nazi moments, to its legacy as a divided city between the soulless plazas of the GDR and western 'colour' – the new development at Potsdamer Platz is an open declaration of Berlin's future as part of the global present. The city's past events and infamous times have been invoked through their absence and the future organized as a global rather than a local possibility.

Arguably, for both Sony and the Daimler Chrysler corporation the all-to-well-known history of Berlin, its discrete events and contrasting moments, had to be discarded in order to display the present. However, in treating the past as one monolithic association, its diverse moments interrupt the present in both mocking and destabilizing ways. The only fully intact building in the whole development that has a claim to the past, for example, is the Haus Huth, a resplendent five-storey building with its original grey stone facade unscathed. Yet in standing apart so clearly from the rest of the modern buildings, it wears its history in a theatrical way, as if it were part of a stage-set in a new consumption play. The idea that it *is* history is lost in the play of modern meaning that surrounds it, where its difference merely serves the overall entertainment function of the complex.

In fact, the same can be said for the Potsdamer Platz Arkaden, a shopping mall which opens next to the Haus Huth and is consciously styled on its nineteenth-century counterpart. Yet in contrast to the earlier arcade, the contemporary Arkaden is no passageway, a visual space of connection between one street and the next that often hid the odds and ends that failed to see the light of day in the then fashionable department store. For the passer-by in today's Arkaden there is no connection to the beyond, only the enticing spaces of the mall itself, designed and laid out like the Sony plaza to script the movements of those who enter and browse. As part of an attempt to forge a seductive presence, the invocation of the past in this instance – in the shape of the arcade – has nothing to do with actually reproducing the past and everything to do with showing how history may function in the present. If it were otherwise, then we would not be talking about a seductive presence but one of a rather different kind.

More or less powerful

It may be helpful to round off this part of the discussion with a reminder that although it is true that power has to have a presence to be effective, the nature of that presence and its effect will vary from mode to mode. Just as there is no everywhere to power, so there is no such thing as a universal blanket presence. Seduction, as we saw in the case of the Potsdamer Platz, exercises a weak presence establishing little more than a series of invitations which risks a minimal impact. Its modest presence, however, belies its potential effectiveness at a distance. For example, Sony's connected presence from afar is experienced through the impression of intimacy and contact, yet all of this may owe little to the corporation actually 'being there'.

In contrast, the kind of authority relations spoken about in the City of London are directly affected by the relationship between proximity and presence. As noted in the previous chapter, because authority relations work through recognition, the more direct the presence, the greater the leverage obtained. Domination, too, because it works through the exercise of near blanket constraints, enhances its presence when it acts among an enclosed rather than a dispersed population. In the case of both domination and authority, however, this is not to imply that the effectiveness of their presence can be measured simply by charting the degree of proximity involved. As I have had cause to stress all along, there is no spatial template for power, so it is not a case of drawing any foregone conclusions about the effectiveness of different modalities over space. It is none the less possible to draw conclusions about imputed presence on the basis of what domination and authority are – given their *particularity* as a relation of power.

It would, of course, be possible to continue in this vein, pointing out the indirect presence that manipulation can achieve, for example, or the decidedly immodest presence that coercion unveils as part of its very nature, but the argument put forward in these pages does not rest upon a loose spatial typology of modes. Rather, it rests, as I have stressed throughout, upon the practical knowledge that power comes in different guises, the effects of which owe much to their diverse geographies of proximity and reach.

What I have tended to assume rather than spell out thus far, however, is that this knowledge also holds for power when it is exercised *with* rather than *over* people.

A Mutual Presence

In chapters 3 and 5, I spoke about power as potentially an exercise in collaboration as much as an instrumental act designed to bend the will of others. The two respective ways of thinking about power, one as a positive, enabling force and the other as a negative, more constraining set of actions, was traced in chapter 3 to rather different understandings of the role of power in society and what exactly an act of power involved. The notion that power may be seen as a facility, a means to get things done that are generally agreed upon, rather than a force directed towards obtaining leverage over others sparked a discussion about what it meant to root power in what Hannah Arendt called 'mutual action'. In the context of thinking about what it is that puts power and us in place, this associational slant on power opens up the possibility of a different series of arrangements from those of a predominantly instrumental nature.

An associational slant on power and place, for example, raises the prospect of the kinds of alliances formed in Arendt's 'public spaces': places where diverse groups of individuals come together to move forward a broadly similar set of aims. The loose, but none the less co-ordinated, actions between environmental and citizens' groups which preceded the protests against the World Trade Organization in Seattle in 1999, the website mobilization that alerted people to the planned protests, and the 'event' itself, which witnessed groups such as Rainforest Action Network and Art and Revolution involved in street protests with dissenting trade union members drawn from the rank and file of America's organized labour, all bear testament to Seattle's 'moment' as a place of collective mobilization (see Jeffrey St Clair's 1999 Seattle diary). Where the Arendt-style association breaks down, however, is with the eruption of violence on the streets of Seattle after the initial success of the protestors in closing down the WTO proceedings.

Acts of violence, you may recall, for Arendt are the very antithesis of power. They limit the possibility of people pooling their collective energies to achieve a shared goal. The violence against commercial property in Seattle by Black Bloc anarchist factions amongst others, the widespread window smashing, the attacks upon TV news crews, as well as the violent street battles with police and the subsequent chaos and street clearances go directly against the grain of Arendt's understanding of power. In place of a vocabulary of negotiation with and persuasion of others, or the exercise of authority amongst them, violence is seen to destroy relationships between people and individualize subjects. This is not to overlook the alliance building between groups that took place prior to the planned protests, much of it at a distance, conducted over the internet; but as an example of associational power, Seattle as a place-based event would fall short of Arendt's collaborative sense of presence. Even apart from the violence, the politics of rejectionism, as Simon Bromley (2001) has characterized it, that Seattle symbolizes does not amount to the kind of collaborative engagement where political solutions are negotiated and differences accepted.

What happened at Seattle, though, would probably have met with Lefebvre's approval: as a spontaneous 'counter-space', or 'counter-presence' (see Lefebvre 1991a: 381–5).

Counter-presence?

Drawing upon the influence of situationist ideas, Lefebvre would very probably have been heartened by the style of protest that characterized Seattle, and indeed by many of the other imaginative events and protests organized of late by the likes of London Greenpeace, Reclaim the Streets and anti-GM protestors in the UK. The ritualized fervour of carnival-style action that has been common to a number of recent protests in the UK in particular, but elsewhere too, has not gone unnoticed,[10] especially in respect of their imaginative use of strategic surprise, symbolic provocation, gestural parades and even ridicule (laughing at the formal authority on display, for instance) in their repertoire of collective action.

Lefebvre would certainly have liked the 'unstaged' effect of the protests and their spontaneous quality, but he would also have endorsed the nature of the spatial challenge involved in a number of those events. In the Carnival Against Capital march in London in June 1999, for instance, one of the central aims of the protest was to undermine the representation of the City of London as an overbearing financial space. The City's rather bloated sense of itself as an important, authoritative space was precisely what the protest hoped to deflate. How effective such a collective mobilization was should be judged less by its fleeting nature and more by the extent to which such symbolic provocation actually unsettled bankers and the like. Lefebvre might well have regarded such a provocative moment as a 'counter-space', inserting:

> itself into spatial reality; against the Eye and the Gaze, against quantity and homogeneity, against power and the arrogance of power, against the endless expansion of the 'private' and of industrial profitability; and against specialised spaces and a narrow localization of function. (1991a: 382)

In the same vein, Lefebvre offers as representative 'counter-spaces' the examples of anti-road campaigns, anti-housing development protests, or

10 Alan Scott and John Street (2000) draw attention to the shared festival style involved in such protests and the role of 'organized spontaneity' in mobilizing collective action. The aestheticization of politics that they document makes explicit reference to the significance of Situationist ideas for contemporary forms of protest and the 'borrowing' of techniques which highlight the embodied, affectual content of action where events are 'seized', so to speak, rather than formally planned.

the demand for spaces which provide an alternative to the prevailing single-minded form. Above all, an alternative means of using space which represents a 'lived' presence is valued: one that, as said before, takes its shape from the routine of its users and their symbolic attachments.

But at this point Lefebvre's analysis seems to fall back on itself (again maybe the outcome of poor editing, although this is possibly generous), leaving us to disentangle what is and what is not an alternative presence. What obscures the analysis, for me at least, is the slippage into oppositional rhetoric: against the arrogance of power, against sameness, against, for that matter, just about anything to do with the 'dominant' coding of space. In short, we are back on the ground of domination and resistance, where mobilization is always against or countering something, not for it in its own right, and power, strangely, is always on the other side of the wire. On this view, power is something that can be identified, pushed back, interrupted, but not something that may arise through collective mobilization on an often loose and tenuous basis. It is taken to be absent, for example, from the mutual action which lies behind what are perhaps best understood as the 'breathing spaces' which make it possible to live differently *among* others more powerful in resource terms.

This is odd, really, especially from someone like Lefebvre who does understand the tangled presences that make up social space. But it is more than merely odd, for such a location of power is also disabling. In the first place, if power is always seen to be from above and 'over there' it becomes impossible to grasp that it may be exercised with rather than against others, and in the chaotic jumble of arrangements we live in, sometimes both at one and the same time.[11] And second, in locating power as elsewhere it becomes difficult to appreciate that the spacing and timing of activities can themselves serve as a resource to be mobilized. Places are, in that sense, more than a series of venues for protest and opposition; they are an *active* part of what it means to put people in place.

In the same way as earlier I probed Lefebvre's account of domination and found rather more to be going on in the name of power, it is likewise possible to recast the oppositional tactics that he describes as a form of associational power in their own right. His sense of those spaces 'appropriated' within the confines of a more powerful presence comes closest to realizing this possibility. The ability to mobilize one's surroundings and to

11 The idea that in the phenomenological flow of everyday life we might find ourselves caught up in multiple relationships of power, subordinate in some relationships whilst superordinate in others, is one that Georg Simmel was at pains to document. Although the idea of exercising power among others does not enter his vocabulary, his sense of our often being at the intersection of various power relationships is a useful reminder of alternatives to the more familiar view of relationships of power as one-dimensional. See Kurt Wolff's *The Sociology of Georg Simmel* (1950) and also Walter (1973).

fashion them in such a way that they reflect an alternative set of codes and attachments is itself an enabling act, one designed perhaps less to counter a dominant presence and more to inhabit it in a positive vein. The negotiations among those actively seeking to create an alternative set of spaces, to use them in ways that are less confrontational and more adaptive, may serve to dilute homogeneous spaces, but the intent may be to carve out a positive-sum scenario where a more beneficial use of space for all is envisaged. The office and factory spaces 'appropriated' in the name of democratic self-management by the Solidarity trade union movement in Poland in the 1980s, first in Gdansk and then across the country, come to mind. So, too, do many of the tactics orchestrated by certain environmental organizations, which give a lead by adopting a form of environmental best practice and then aiming to replicate such 'lived' practices through the art of persuasion (and non-violent protest) in the hope of their becoming routinized, as part and parcel of a more imaginative, sustainable presence in communities.

Neither democratic self-management in Poland nor the space for environmental sustainability is much in evidence today. The former in fact has all but disappeared, whilst hope for the latter remains elusive. My interest in them rests upon the idea of an alternative set of arrangements where power is present as a collaborative force, as empowering of those taking part, rather than simply as something to rally against. In that respect, what perhaps Lefebvre could have learnt from Arendt is that being placed among others can be a positive as well as a negative experience.

Placing Power

The main point to be drawn from this chapter is, I hope, the by now familiar one that it is the very nature and constitution of places themselves – their different mix of rhythms, routine practices and attachments – which gives rise to *tangled* arrangements of power. Places are made up of the cross-cutting nature of people's lives as they go about their daily business, and the manner in which power is exercised across such colliding spaces is what gives them much of their shape and character. It is also what gives those who inhabit them the often *contradictory experience of remoteness and proximity* in social relationships. In line with the differences between places, therefore, power may be seen to mutate in ways that, for example, see acts of domination closing down access to routine service-workers in their own place of employment, alongside the manipulative concealment of their choices; locals and passers-by physically barred from a gated housing estate, a bar reinforced by the threat of coercion; visitors and shoppers enticed to move through the public spaces of a shopping mall in ways that they might not otherwise have done, sweetened by rewards for compliance;

or a 'breathing space' negotiated by anti-road campaigners, sealed by authority exercised among rather than over others. Above all, I want to stress that places may play host to different groups and institutions marking their presence in any variety of cross-cutting arrangements, where the past as an active trace may also serve to entangle the present.

At the very least, I hope that I have been able to show that if power has a presence at all, then that power may be modest, limited to suggestion and enticement, or experienced as a series of almost total blanket conditions, or felt directly through the obvious lines of authority, or even experienced indirectly through covert, less visible means. As I have stressed, the question of what relations of power are in place remains an empirical one, open to investigation of the spacing and timing of people's interactions; some of which will involve the actions of those co-present, in real time, whereas others will involve the actions of those nearby, in close spatial proximity.

But whatever mixture of distanciated and proximate actions is involved, the *effects* of such mediated actions are always felt at first hand. People are placed by power; they experience it through the rhythms and relationships of particular places, not as some pre-packaged force transmitted intact from half-way across the globe. It comes back to my oft-repeated comment that power neither travels nor flows, whereas resources do, which once accepted allows us to think about power as inseparable from its effects. Understanding that such effects may emanate from afar, that power has reach, simply means that their presence is more or less mediated in space and time, but – as I have insisted – it is the various points of power's application which *place* us on the receiving end and which often have the ability to make us feel remote and distant from others.

8

Conclusion: Misplaced Power

Hopefully, what I have said up to now is sufficient to show why I think *Lost Geographies of Power* to be an apt title. Even though it would be misleading of me to suggest that the spatial aspects of power have faded from view in much of the power literature, I do wish to press the point that much of what we take to be the spatial trappings of power has obscured rather more than it has revealed about the workings of power.

On the one hand, I have been trying to show that despite a richer spatial vocabulary of power than hitherto, we seem to have lost our way in a rather generalized landscape of power, where its traces, we are told, are there for all to see, if only we look hard enough. For some, it is the circulation and flow of power which catch their imagination, usually in durable networks but sometimes simply in free-flow; for others it is the dispersed arrangements of power which offer a purchase on what it means to bring people within reach, exercised in immanent fashion or otherwise; and for yet others it is the everyday spaces of power which convey something of what it means to be immersed in the constitutive detail that puts each and every one of us in place. Yet, from whatever angle you look at it, there is nothing in the landscape that reminds you that power relations have long been experienced through a variety of different modes and that the likes of domination, authority, manipulation and seduction are always already spatial. It is power's diverse geographies that surely have been lost.

On the other hand, I am not altogether convinced that anything really was ever truly lost; misplaced, in terms of understanding, but perhaps not actually lost in the sense that the geographies of power were out there waiting to be found. Being a little more curious about what it is that makes, for example, authority or seduction peculiarly distinctive spatial acts has possibly shown us something that we knew all along – although only now do we understand it. We may, for instance, have known about seduction's modest and dilute qualities through habit, routine and our exposure to advertising, where the possibility of refusal or indifference is evident, yet we may never have given any real thought to the fact that those self-same qualities lie behind its potential arm's-length reach. Indeed, why should we have given such matters any thought? Unless the question is raised, its significance remains irrelevant to us, a step too far in terms of our understanding.

Bearing this in mind, my own sense is that we have not so much lost sight of all that is interesting about power and the ways in which it is exercised as settled for a more familiar, albeit misplaced, understanding. If I have found anything in the course of writing this book, it is an understanding of power that has long been with us, yet never actually been taken apart to reveal its diverse geographies of proximity and reach.

Much of this book has in fact been given over to reminding us about the particularities of power, its specific modal qualities. But this reminder has also served to prompt curiosity about how particular kinds of power take effect close to, nearby, at a distance and through the tangled arrangements of place. It is not enough to accept blithely that the likes of domination or authority, or indeed both, will be achieved through political or economic bodies simply exercising 'command' over space and time. Reach, proximity and presence are *not* givens; they make a difference to the exercise of power precisely because the many and varied modalities of power are themselves constituted *differently* in space and time. Power, I have argued, is inseparable from its effects; it does not reach out over space prior to having an impact. There is no uniform substance called power that has the (enchanted) ability to circumnavigate the next block, let alone the globe. Resources and their packaged capabilities may do so, but not the mediated effects of power.

Yet, even if we believe that, there is no guarantee that the impression will be a lasting one, should we find ourselves constrained by the mindless act of a remote so-called authority or deliberately deceived by a rogue corporation with its headquarters on the other side of the planet. It is easy to be lost in familiar spaces. And it is just as easy to slide between power, domination and authority as if they amounted to much the same thing in terms of their impact upon our lives. Power and domination, in particular, trip off the tongue as synonyms; the terms can be switched with relative ease with, it would seem, no apparent loss of meaning.

Now we cannot get around any of this by pretending that it does not happen, but we can draw attention to its consequences, which is what I have tried to do at various stages in the book.

In the first part of the book, I attempted to show that if you think about power as a 'thing-like' property capable of extensive reach in a rather uncomplicated fashion, or as something which is generated through networked interactions and projected unproblematically across their length and breadth, it becomes straightforward to believe that distance is something easily overcome and presence effortlessly established. Rigid distances are either traversed by power being redistributed upwards or downwards in a linear fashion, or overcome in a fluid manner as power is organized, channelled and transmitted around the networks to achieve far-flung goals. Equally, if you think of power as a rather undifferentiated quality, constituted through a multitude of practices which in true immanent fashion serve to govern our conduct, then it is easy to imagine how government of each and everyone achieves its totalizing reach. The habituation of the mind and body that follows seems to require an apparatus of rule that needs little, if any, discrete spatialization.

In the second part of the book, I have sought not so much to minimize or ridicule these views as to turn what is useful about them to new purposes. What we have lost in this rather generalized landscape of power, where distances are often fixed and unyielding and proximities established and settled, I have sought to recover by problematizing both power and its geography.

In the case of power's characteristics and qualities, I have tried to live up to the legacies of Max Weber and Hannah Arendt by taking back their understanding that power is always power of a particular kind. But in doing so, I have recast this understanding in a topological setting where the diverse geographies of power's proximity and reach are no longer so rigid. Equally, I have tried to retain the sense in which Foucault understood power as an immanent force coextensive with its field of operation, yet within a framework that acknowledges the time-honoured particularities of power: that the constraining effects of domination are neither the same as the unnerving threats of coercion, nor similar to the suggestive qualities of seduction or the conceded act of authority, and so on. The result, I hope, is a different and largely renewed understanding of the ways in which geography and power exercise us.

Topological Findings

I concluded my remarks on Foucault and Deleuze's immanent conception of power in chapter 4 by saying that, somewhat surprisingly, spatiality as a focus had faded from view the further that we moved away from consideration of detailed institutional settings. Once the confined arrangements of the prison or the clinic or the military barracks gave way to the dispersed arrangements of government, there seemed to be less curiosity about how a diverse and scattered population could be drawn within reach. As I see it, Foucault and Deleuze's topological understanding of power became all the weaker for the evacuation of the spatial.

The notion of topology, it should be said, is a broad one. Within geographical studies, the term has recently attracted attention for, among other things, its contrast to a view of the world comprised of so many scales or levels of social activity – from the local and the regional through to the national and the global (see, for example, Amin, 2002; Bingham and Thrift, 2000; Hinchliffe, 2000; Murdoch, 1998; Ogborn, 1998; Thrift, 1999). One of the attractions of a topological view of social relations is that it diverts attention away from the geographical scale at which, say, corporate bodies or political forces are supposed to operate and focuses attention instead upon the relational arrangement of which they are a part. In particular, a topology of social relations should help us to focus upon their co-constitutive nature *and* the spaces and times they actively construct in the process.

This may sound a little abstract, but all that it really means is that there are no pre-defined distances or simple proximities to speak about in relation to an exercise of something like corporate or state rule. Domination in real time, for instance, may establish an immediate presence in far-off locations and, likewise, seduction may lend itself to simultaneous interaction through near instantaneous technologies, yet their effects may play across one another in the *same* place as part of a cross-cutting arrangement of power. In distance terms, what is near and what is far is not simply a question of geometric measurement between fixed points; rather it is one of connection and simultaneity as different groups and institutions mark their presence through interaction in all kinds of powerful and not so powerful ways. It is this topological slant that I think best serves my purposes here, rather than those interpretations which stress the unchanged nature of entities under spatial distortion (see Law 1999).

There are two aspects of this that have shaped my thinking and directly influenced my treatment of power and spatiality.

First, in my attempt to develop a topology of power sensitive to the diverse geographies of proximity and reach, I worked directly from the subtle distinctions of power, rather than from any generalized sweep of relations. Thus the particular way in which something like authority takes effect, for example, by working through relations of proximity and presence was shown to differ from the indirect reach that characterizes a one-side relation like manipulation. The difference in their mode of operation is, I would argue, critical because, as the example of the big commercial houses in chapter 6 illustrated, the recognition of their 'global' authority means everything to them, even if they also engage in manipulative activities to maintain their global position. Whereas both may be exercised at a distance, either through a succession of mediated relationships or by the establishment of an immediate presence through real-time technologies, for a corporate authority to have any leverage its presence requires proximity.

And this scope is achieved not by such agencies shifting scale by moving back and forth between the local and the global, but rather by a loosening of defined distances and times to establish a more proximate authority. It is in this sense that the far-off is brought into reach for such 'authoritative' bodies.

If the ability to draw distant others within close reach is one side of the topological equation, the other that I have attempted to elaborate is the ability of groups and institutions to make the close-at-hand seem surprisingly distant. In chapter 7, I approached this second aspect by considering the ways in which people move in and around one another, sliding across each other's lives in a more or less powerful fashion to make their presence felt. In this topology of places, what is near and what is far, who 'belongs' and who does not, may be distorted by the placement of forces and their relational ties, where the presence of others may be smothered, excluded, barred, threatened, enticed or simply constrained. My point, however, was not to show that all places are saturated with power, but rather to suggest that their tangled arrangements of power may produce a degree of remoteness and proximity in social relationships that owes much to other times and other places also being present. It is, for me, the surprising juxtapositions that arrangements of power throw up which suggests that there are no simple proximities.

These two findings hardly exhaust what there is to say about topological arrangements of power, but to my mind they do at least point to where a more geographically curious dialogue of power could go. Side-stepping approaches that assume fixed distances, settled proximities, unproblematic extensions, linear projections and fluid transmissions as well as the totalizing reach of power is a start, nevertheless. For my part, I have tried to think through the sense in which power is inherently spatial and, indeed, the ways in which spatiality is itself imbued with power. But I have done so not simply because there is, today, a keener interest in matters of space and spatiality, but rather because it *matters* that we understand power in all its guises – what it is that exercises us and how – as well as why it seems to turn up in such an array of places. It matters because if we fail to recognize the diverse ways in which power puts us in place, we disempower ourselves.

So, I want to end on this rather pointed note: why it *matters* that we do not lose ourselves in the familiar spaces of power.

The Whereabouts of Power

Let me start with a declaration: the durable architecture of power that I have spoken about is not likely to go away and there would appear to be little point in wishing it would. The (false) impression that power is something available to be passed on to those responsible enough to take decisions, or the view that the weighty institutions of this world 'have' the power to put us in our place, are almost too obvious to bear consideration. One of the reasons, to my mind, for this enduring quality is the ability of 'power' to make itself visible when it so requires, or rather it is the ability of those who exercise it to mobilize its material representation successfully. Grand buildings, imposing monuments, high walls, gated communities, and aircraft carriers that blot out the horizon help in this respect. The sheer concentration of decision-makers, money or might, or perhaps all three, at a particular point or location appears to tell us where power ultimately lies and who has it. It also gives us something to resist, to defy, and even to run away from should the odds be stacked against us.

Such *centred* spaces, if that is the right description, are real enough, but they are not 'filled' with power; rather they are, as I have stressed throughout, resource bases that have been mobilized over long or short periods of time. As argued in chapter 5, power, unlike resources, is not something fixed in monumental buildings, state ministries, fenced-off housing estates, or the headquarters of global multinationals for that matter. It may appear that way, but the impression is misplaced. Equally, however, you cannot 'dissolve' such resource bases into a networked morphology of flows, inasmuch as the territorial bundling of resources rightfully may inspire trepidation and fear at the moment of confrontation. But, and this is the important point, it is not resources but the exercise of power which actually puts us in place. Resources and abilities, as we know, may be misused, lost, wasted or used to nil effect. That does not necessarily lessen their threatened use, of course, but if this durable architecture is all that we recognize as power, we are likely to miss most of what exercises us.

At the non-durable end of power, if I may put it like that, is the *decentred* view that power, far from being located at the apex of anything, is an immanent, normalizing force that works through people's lives, shaping their very being in a way that defies spatial definition. I have drawn the contrast sharply not to caricature this admittedly rather scary view of power, but to paint it at the extreme. This is not Foucault's non-subjective but intentional, coextensive but immanent, portrait of power. But it does share with Foucaldian-style accounts an assessment of power that is judged by its intended effects rather than by its actual effects. The all encompassing and individualizing apparatus of rule described by Hardt and Negri (2000), where the production of new subjectivities is assumed to take its shape from the simple act of living, has, as I have argued, a ring of hollowness about it. There is an emptiness precisely where the spatial and temporal mediations of power should be. Thinking something is likely to happen is not at all the same as knowing, or at least being curious about, how the lives of distant others are brought within reach.

If we believe that power works in such ill-defined and unformed ways then, on this account, not only would we fail to recognize that operating a remote-controlled electronic gate at the entrance to an exclusive residential estate is the modern counterpart of raising the drawbridge, we would also fail to grasp the more dispersed, mediated arrangements of power that exercise us. In particular, we would be unaware of the diverse ways in which geography shapes the mediated presence of power's many and varied relationships.

It seems to me that if we are forced to choose between either a centred or a decentred view of power, or to shuffle between them in an effort to blur clearly demarcated scales and boundaries, geography is the loser. At the durable end, the impression given would be one of resourced capabilities remaining on hold and, once triggered, spreading outwards from an identifiable centre to realize their potential. The sense in which a centre can be said to possess capabilities, as we saw in chapter 2, presupposes defined distances and unproblematic extensions in a way that reduces geography to a series of footnoted relations. Power and space become separable, and the latter operates more or less as an add-on. In contrast, where there are no 'centres' of power to speak of, no defined territorial spaces to occupy or administer, power and space become one. They are inseparable, but only in the sense that they fill out what is there because nothing is outside the seamless logic of rule. Power, as Naomi Klein (2001b) colourfully suggests, is so everywhere, it seems nowhere. It is placeless, devoid of any geography, let alone the kind that lends itself to easy mapping.

What pass unrecognized in all of this, of course, are the diverse geographies of power that I have outlined: the topological arrangements that suppose neither defined distances nor ill-defined social spaces. But to appreciate this does not mean that we have to avoid, even if we could, all talk of centralized resources and territorially embedded capabilities or the dispersed and diffuse effects of the exercise of power. It is just that a centred/decentred binary is not particularly helpful in drawing attention to the mediated arrangements of power that exercise us. The fantasy of power as something which is *either* centred *or* decentred allows us to believe that, if it does not overstretch itself, it remains a force to be reckoned with, or, alternatively, if it is a force so pervasive and all encompassing in its reach it leaves little room for manoeuvre. Either way, the effect is potentially disarming.

If power is thought of as centred in an obviously tangible way, whether in the shape of some 'sleeping' biotechnology giant or the sheer concentration of Hollywood's cultural capabilities, the identifiable whereabouts of power invites resistance to it. At worst, the powerful and the powerless are lined up in some assumed confrontational formation, where the latter's abilities to overcome the 'greater' power are always in doubt. Interruption, subversion and direct opposition are all possible, but to think of succeeding in such an unequal power play seems almost to tempt providence. Thus it is easy for a degree of fatalism to enter into the proceedings, as events, in true Barry Hindess (1996) reasoning, pan out in favour of the powerful. Likewise, if power is believed to have no real whereabouts to speak of, no obvious landmarks to target, pitting oneself against it is a tricky, if not hopeless, option.

If we abandon these admittedly rather simplistic yet surprisingly familiar formulations of power and resistance, and concentrate instead on the mediated arrangements of power spelt out in this book, a more provisional, less daunting view of power opens up. In the kinds of articulated arrangements that I have endeavoured to show, power often makes its presence felt through a variety of modes playing across one another. The erosion of choice, the closure of possibilities, the manipulation of outcomes, the threat of force, the assent of authority or the inviting gestures of a seductive presence, and the combinations thereof, are among the diverse ways in which power puts us in place. A simple domination/resistance framework in this respect trivializes the feeling for what power is when it is brushed up against. It matters, as I have argued throughout this book, that the many and varied modalities of power are recognized as constituted differently in space and time. Otherwise, we lose sight not only of the mediated nature of power, but also of what it is that confronts us. Before we can embark upon alternative paths to action and social change, we need to be aware of what it is that we face and how power in its more provisional yet spatially nuanced guises exercises us.

We lose sight too of what it means to engage in *associational* arrangements of power in response to those who seek only instrumental advantage. In the former, people are not powerless; they exercise power through the positive strengths of collaborative association. Different modes of empowerment enable people to come together to pursue a common end, without having to fit themselves into the resistance mould of politics. Negotiation and shared outcomes replace confrontation and opposition and take us into the realm of the 'power to' act, rather than the domain of the 'powerless' who are likely to be left feeling that 'power over' them is all that they are every likely to experience.

Which brings me full circle in some ways, for I started this book by musing about the fact that the association of power with geography is a familiar one, yet more misleading than revealing about what it can tell us about the difference that geography makes to the exercise of power. It says something about the hold that the language of 'power over' others has on our spatial imaginations that the very idea of empowerment remained largely to one side of my deliberations. In bringing the differences between particular kinds of instrumental power to the fore and weaving them in a more geographically curious dialogue, it is perhaps time to do likewise for the *transverse* modes of powers that Hannah Arendt in particular identified: those that involve the exercise of power *with* rather than over others. Whilst I touched upon the presence of such mutual action in chapter 7, I think that we have lost sight of the whereabouts of this altogether different guise of power too. Perhaps that is another geography lost waiting to be found, or rather, I should say, waiting to be understood for the first time.

Bibliography

Agnew, J. (1994) 'The territorial trap: the geographical assumptions of international relations theory', *Review of International Political Economy* 1(1), 53–80.

Agnew, J. (1998) *Geopolitics: Revisioning World Politics*, London and New York: Routledge.

Agnew, J. (1999) 'Mapping political power beyond state boundaries: territory, identity, and movement in world politics', *Millennium* 28(3), 499–521.

Agnew, J. and Corbridge, S. (1995) *Mastering Space: Hegemony, Territory and International Political Economy*, London and New York: Routledge.

Aksoy, A. and Robins, K. (1992) 'Hollywood for the 21st century: global competition for critical mass in image markets', *Cambridge Journal of Economics* 16(1), 1–22.

Allen, J. (1997) 'Economies of power and space' in Lee, R. and Wills, J. (eds) *Geographies of Economies*, London and New York: Arnold, 59–70.

Allen, J. (1999a) 'Spatial assemblages of power' in Massey, D., Allen, J. and Sarre, P. (eds) *Human Geography Today*, Cambridge: Polity, 194–218.

Allen, J. (1999b) 'Cities of power and influence: settled formations' in Allen, J., Massey D. and Pryke, M. (eds) *Unsettling Cities: Movement and Settlement*, London and New York: Routledge/ Open University, 181–218.

Allen, J. (2000a) 'Power: its institutional guises (and disguises)' in Hughes, G. and Fergusson, R. (eds) *Ordering Lives: Family, Work and Welfare*, London and New York: Routledge/ Open University, 7–43.

Allen, J. (2000b) 'On Georg Simmel: proximity, distance and movement' in Crang, M. and Thrift, N. (eds) *Thinking Space*, London and New York: Routledge, 54–70.

Allen, J. (2002) 'Power' in Agnew, J., Mitchell, K. and Ó Tuathail, G. (eds) *A Companion to Political Geography*, Oxford: Blackwell.

Allen, J. and Pryke, M. (1994) 'The production of service space', *Environment and Planning D: Society and Space* 12(4), 453–75.

Amin, A. (2002) 'Spatialities of globalisation', *Environment and Planning A* 34(3), 385–99.

Arendt, H. (1951) *The Origins of Totalitarianism*, New York: Harcourt Brace.
Arendt, H. (1958) *The Human Condition*, Chicago and London: University of Chicago Press.
Arendt, H. (1961) *Between Past and Future: Six Exercises in Political Thought*, London: Faber and Faber.
Arendt, H. (1970) *On Violence*, San Diego: Harvest.
Aron, R. (1986) '*Macht*, power, *puissance*: democratic prose or demoniacal poetry?' in Lukes, S. (ed.) *Power*, Oxford: Blackwell, 253–77.
Barnes, B. (1988) *The Nature of Power*, Oxford: Blackwell.
Barnett, C. (1999) 'Culture, government and spatiality: re-assessing the "Foucault effect" in cultural-policy studies', *International Journal of Cultural Studies* 2(3), 369–97.
Beaverstock, J. V., Smith, R. G. and Taylor, P. J. (1999a) 'A roster of world cities', *Cities* 16(6), 445–58.
Beaverstock, J. V., Smith, R. G. and Taylor, P. J. (1999b) 'The long arm of the law: London's law firms in a globalising world economy', *Environment and Planning A* 31(10), 1857–76.
Beaverstock, J. V., Smith, R. G. and Taylor, P. J. (2000a) 'Globalisation and world cities: some measurement methodologies', *Applied Geography* 20, 43–63.
Beaverstock, J. V., Smith, R. G. and Taylor, P. J. (2000b) 'World-city network: a new metageography?', *Annals of the Association of American Geographers* 90(1), 123–34.
Benhabib, S. (1992) *Situating the Self: Gender, Community and Postmodernism in Contemporary Ethics*, Cambridge: Polity.
Benhabib, S. (1996) *The Reluctant Modernism of Hannah Arendt*, London and Thousand Oaks, CA: Sage.
Benton, T. (1981) 'Objective interests and the objectivity of power', *Sociology* 15, 161–84.
Bhaskar, R. (1975) *A Realist Theory of Science*, Leeds: Leeds Books.
Bhaskar, R. (1979) *The Possibility of Naturalism: A Philosophical Critique of the Contemporary Human Sciences*, Brighton: Harvester.
Bhaskar, R. (1986) *Scientific Realism and Human Emancipation*, London: Verso.
Bhaskar, R. (1989) *Reclaiming Reality: A Critical Introduction to Contemporary Philosophy*, London: Verso.
Bhaskar, R. (1993) *Dialectic: The Pulse of Freedom*, London: Verso.
Bhaskar, R. (1994) *Plato, Etc.: The Problems of Philosophy and their Resolution*, London: Verso.
Bingham, N. and Thrift, N. (2000) 'Some new instructions for travellers: the geography of Bruno Latour and Michel Serres' in Crang, M. and Thrift, N. (eds) *Thinking Space*, London and New York: Routledge, 281–301.
Boden, D. and Molotch, H. L. (1994) 'The compulsion of proximity' in Friedland, R. and Boden, D. (eds) *NowHere: Space, Time and Modernity*, Berkeley, Los Angeles and London: University of California Press, 257–86.
Bourdieu, P. (1989) 'Social space and symbolic power', *Sociological Theory* 7(1), 14–25.
Bourdieu, P. (1991) *Language and Symbolic Power* (ed. J. B. Thompson, trans. G. Raymond and M. Adamson), Oxford: Polity.

Brenner, N. (1998) 'Between fixity and motion: accumulation, territorial reorganization and the historical geography of spatial scales', *Environment and Planning D: Society and Space* 16(4), 379–504.
Bromley, S. (2001) 'The golden straitjacket: moving on from Seattle', *Radical Philosophy* 107, 5–10.
Burchell, G., Gordon, G. and Miller, P. (eds) (1991) *The Foucault Effect: Studies in Governmentality*, Chicago: University of Chicago Press.
Caldeira, T. (1996) 'Fortified enclaves: the new urban segregation', *Public Cultures* 8, 303–28.
Callon, M. (1986) 'Some elements of a sociology of translation: domestication of the scallops and the fishermen of St Brieuc Bay' in Law, J. (ed.) *Power, Action, and Belief: A New Sociology of Knowledge?*, London, Boston and Henley: Routledge and Kegan Paul, 196–233.
Callon, M. and Latour, B. (1981) 'Unscrewing the Big Leviathan: how actors macro-structure reality and how sociologists help them to do so' in Knorr-Cetina, K. and Cicourel, A. (eds) *Advances in Social Theory and Methodology: Towards an Integration of Micro- and Macro-Sociologies*, Boston, London and Henley: Routledge and Kegan Paul, 277–303.
Castells, M. (1996) *The Rise of the Network Society*, Oxford: Blackwell.
Castells, M. (2000) 'Materials for an explanatory theory of the network society', *British Journal of Sociology* 51(1), 5–24.
Clarke, J. and Newman, J. (1997) *The Managerial State: Power, Politics and Ideology in the Remaking of Social Welfare*, London and Thousand Oaks, CA: Sage.
Cochrane, A. and Jonas, A. (1999) 'Reimagining Berlin: world city, national capital or ordinary place?', *European Urban and Regional Studies* 6, 145–64.
Cooper, D. (1998) *Governing Out of Order: Space, Law and the Politics of Belonging*, London and New York: Rivers Oram Press.
Crang, M. (2001) 'Rhythms of the city: temporalised space and motion' in May, J. and Thrift, N. (eds) *Timespace: Geographies of Temporality*, London and New York: Routledge, 187–207.
Crang, M. and Travlou, P. S. (2001) 'The city and topologies of memory', *Environment and Planning D: Society and Space* 19(2), 161–77.
Dahl, R. A. (1957) 'The concept of power', *Behavioural Science* 2, 201–15.
Dahl, R. A. (1961) *Who Governs?*, New Haven, CT: Yale University Press.
Dalby, S. (1991) 'Critical geopolitics: discourse, difference, and dissent', *Environment and Planning D: Society and Space* 9(3), 261–83.
Dalby, S. and Ó Tuathail, G. (1998) (eds) *Rethinking Geopolitics*, London: Routledge.
Dean, M. (1996a) 'Foucault, government and the enfolding of authority' in Barry, A., Osborne, T. and Rose, N. (eds) *Foucault and Political Reason: Liberalism, Neoliberalism and Rationalities of Government*, London: UCL Press, 209–29.
Dean, M. (1996b) 'Putting the technological into government', *History of the Human Sciences* 9(3), 47–68.
Dean, M. (1998) 'Questions of method' in Velody, I. and Williams, R. (eds) *The Politics of Constructionism*, London and Thousand Oaks, CA: Sage, 182–99.

Dean, M. (1999) *Governmentality: Power and Rule in Modern Society*, London and Thousand Oaks, CA: Sage.
Deleuze, G. (1988) *Foucault* (trans. S. Head), London: Athlone Press.
Deleuze, G. (1995) *Negotiations* (trans. M Joughin), New York: Columbia University Press.
Deleuze, G. and Guattari, F. (1988) *A Thousand Plateaus: Capitalism and Schizophrenia*, London: Athlone Press.
Deleuze, G. and Parnet, C. (1987) *Dialogues*, London: Athlone Press.
Disch, L. J. (1994) *Hannah Arendt and the Limits of Philosophy*, Ithaca, NY, and London: Cornell University Press.
Dodds, K. (1993) 'Geopolitics, experts and the making of foreign policy', *Area* 25, 70–4.
Dodds, K. (1994) 'Geopolitics in the Foreign Office: British representations of Argentina 1945–1961', *Transactions of the Institute of British Geographers* 19, 273–90.
Dodds, K. (2000) *Geopolitics in a Changing World*, Harlow: Prentice Hall.
Dodds, K. and Sidaway, J. D. (1994) 'Locating critical geopolitics', *Environment and Planning D: Society and Space* 12 (5), 514–24.
Driver, F. (1985) 'Power, space and the body: a critical assessment of Foucault's *Discipline and Punish*', *Enivronment and Planning D: Society and Space* 3 (4), 425–46.
Driver, F. (1993) 'Bodies in space: Foucault's account of disciplinary power' in Jones, C. and Porter, R. (eds) *Reassessing Foucault: Power, Medicine and the Body*, London and New York: Routledge, 113–31.
du Gay, P. (1996) *Consumption and Identity at Work*, London and Thousand Oaks, CA: Sage.
du Gay, P. (2000) *In Praise of Bureaucracy*, London and Thousand Oaks, CA: Sage.
Elden, S. (2001) *Mapping the Present: Heidegger, Foucault and the Project of a Spatial History*, London and New York: Continuum.
Foucault, M. (1977) *Discipline and Punish: The Birth of the Prison* (trans. A. Sheridan), London: Allen Lane.
Foucault, M. (1980) *Power/Knowledge* (ed. C. Gordon), Brighton: Harvester.
Foucault, M. (1982) 'The subject and power', in Dreyfus, H. L. and Rabinow, P. (eds), *Michel Foucault: Beyond Structuralism and Hermeneutics*, Brighton: Harvester, 208–26.
Foucault, M. (1984) *The History of Sexuality, Vol. 1: An Introduction*, Harmondsworth: Penguin.
Foucault, M. (1986) *The History of Sexuality, Vol. 3: The Care of the Self*, Harmondsworth: Penguin.
Foucault, M. (1988a) 'On power' in Kritzman, L. D. (ed.) *Michel Foucault: Politics, Philosophy, Culture*, London and New York: Routledge, Chapman and Hall, 96–109.
Foucault, M. (1988b) 'The ethic of care for the self as a practice of freedom' in Bernauer, J. and Rasmussen, D. (eds), *The Final Foucault*, Boston, MA: MIT Press.
Foucault, M. (1988c) 'Technologies of the self', in Martin, L. H., Gutman, H. and Hutton, P. H. (eds) *Technologies of the Self*, London: Tavistock, 16–49.

Foucault, M. (1991) 'Governmentality' in Burchell, G., Gordon, C. and Miller, P. (eds) *The Foucault Effect: Studies in Governmentality*, Hemel Hempstead: Harvester Wheatsheaf, 87–104.

Foucault, M. (2001a) '"Omnes et singulatim": toward a critique of political reason', in Foucault, M. *Power, Essential Works, Vol. 3* (ed. J. D. Faubion, trans. R. Hurley), Harmondsworth: Allen Lane, 298–325.

Foucault, M. (2001b) 'Space, knowledge and power', in Foucault, M. *Power, Essential Works, Vol. 3* (ed. J. D. Faubion, trans. R. Hurley), Harmondsworth: Allen Lane, 349–64.

Giddens, A. (1977) *Studies in Social and Political Theory*, London: Hutchinson.

Giddens, A. (1979) *Central Problems in Social Theory: Action, Structure and Contradiction in Social Analysis*, Basingstoke: Macmillan.

Giddens, A. (1981) *A Contemporary Critique of Historical Materialism*, London and Basingstoke: Macmillan.

Giddens, A. (1984) *The Constitution of Society: Outline of a Theory of Structuration*, Cambridge: Polity.

Giddens, A. (1985) *The Nation State and Violence*, Cambridge: Polity.

Giddens, A. (1990) *The Consequences of Modernity*, Cambridge: Polity.

Giddens, A. (1994) 'Living in a post-traditional society' in Beck. U., Giddens, A. and. Lash, S. (eds), *Reflexive Modernization: Politics, Tradition and Aesthetics in the Modern Social Order*, Cambridge: Polity, 56–109.

Gordon, C. (1987) 'The soul of the citizen: Max Weber and Michel Foucault on rationality and government' in Whimster, S. and Lash, S. (eds) *Max Weber, Rationality and Modernity*, London: Allen and Unwin, 293–316.

Gordon, C. (1991) 'Governmental rationality: an introduction' in Burchell, G., Gordon, C. and Miller, P. (eds) *The Foucault Effect: Studies in Governmentality*, 1–51.

Gordon, C. (2001) 'Introduction' in Foucault, M. *Power, Essential Works, Vol. 3* (ed. J. D. Faubion, trans. R. Hurley), Harmondsworth: Allen Lane, xi–xli.

Graham, S. (1998a) 'Spaces of surveillant simulation: new technologies, digital representations and material geographies', *Environment and Planning D: Society and Space* 16(4), 483–504.

Graham, S. (1998b) 'The end of geography or the explosion of place? Conceptualizing space, place and information technology', *Progress in Human Geography* 22(2), 165–85.

Gregory, D. (1994) *Geographical Imaginations*, Oxford: Blackwell.

Gregory, D. (1995) 'Imaginative geographies', *Progress in Human Geography* 19(4), 447–85.

Gregory, D. (1998) 'Power, knowledge and geography', *Geographische Zeitschrift* 86(2), 70–93.

Habermas, J. (1977) 'Hannah Arendt's communications concept of power', *Social Research*, Spring, 3–24.

Habermas, J. (1989) *The Structural Transformation of the Public sphere: An Inquiry into a Category of Bourgeois Society*, Cambridge: Polity.

Hacking, I. (1986) 'Making up people', in Heller, T. C. et al. (eds), *Reconstructing Individualism*, Stanford: Stanford University Press, 222–36.

Hardt, M. and Negri, A. (2000) *Empire*, Cambridge, MA: Harvard University Press.

Harré, R. and Madden, F. H. (1975) *Causal Powers: A Theory of Natural Necessity*, Oxford: Blackwell.

Hartsock, N. (1985) *Money, Sex and Power*, Boston: Northeastern University Press..

Harvey, D. (1982) *The Limits to Capital*, Oxford: Blackwell.

Harvey, D. (1989) *The Condition of Postmodernity*, Oxford: Blackwell.

Hetherington, K. and Law, J. (2000) 'Theme issue: after networks', *Environment and Planning D: Society and Space* 18(2), 127–32.

Hinchliffe, S. (2000) 'Entangled humans: specifying powers and their spatialities' in Sharpe, J. P., Routledge, P., Philo, C. and Paddison, R. (eds) *Entanglements of Power: Geographies of Domination/Resistance*, London and New York: Routledge, 219–37.

Hindess, B. (1996) *Discourses of Power: From Hobbes to Foucault*, Oxford: Blackwell.

Howe, N. (1998) 'Berlin Mitte', *Dissent*, Winter, 71–81.

Howell, P. (1993) 'Public space and the public sphere: political theory and the historical geography of modernity', *Environment and Planning D: Society and Space* 11(3), 303–22.

Hsing, Y.-T. (1998) *Making Capitalism in China: The Taiwan Connection*, New York and Oxford: Oxford University Press.

Huyssen, A. (1997) 'The voids of Berlin', *Critical Inquiry* 24, 57–81.

Isaac, J. C. (1987) *Power and Marxist Theory: A Realist View*, Ithaca, NY, and London: Cornell University Press.

Jessop, B. (2000) 'The crisis of the national spatio-temporal fix and the tendential ecological dominance of globalizing capitalism', *International Journal of Urban and Regional Research* 24 (2), 323–60.

Kern, S. (1983) *The Culture of Time and Space 1880–1918*, Cambridge, MA: Harvard University Press.

Klein, N. (2001a) *No Logo*, London: Flamingo.

Klein, N. (2001b) 'Reclaiming the commons', *New Left Review* 9, 81–9.

Kracauer, S. (1995) *The Mass Ornament: Weimar Essays* (ed. and trans. T. Y. Levin), Cambridge, MA: Harvard University Press.

Latour, B. (1986) 'The powers of association' in Law, J. (ed.), *Power, Action and Belief: A New Sociology of Knowledge?*, London, Boston and Henley: Routledge and Kegan Paul, 264–80.

Latour, B. (1987) *Science in Action*, Cambridge, MA: Harvard University Press.

Latour, B. (1988) *The Pasteurization of France* (trans. A. Sheriden and J. Law), Cambridge, MA: Harvard University Press.

Latour, B. (1991) 'Technology is society made durable' in Law, J. (ed.) *A Sociology of Monsters: Essays on Power, Technology and Domination*, London and New York: Routledge, 103–31.

Latour, B. (1993) *We Have Never Been Modern* (trans. C. Potter), Hemel Hempstead: Harvester Wheatsheaf.

Latour, B. (1996) *Aramis: Or the Love of Technology*, Cambridge, MA: Harvard University Press.

Latour, B. (1999a) *Pandora's Hope: Essays on the Reality of Science Studies*, Cambridge, MA: Harvard University Press.
Latour, B. (1999b) 'On recalling ANT' in Law, J. and Hassard, J. (eds) *Actor Network Theory and After*, Oxford: Blackwell, 15–25.
Law, J. (1986a) 'On the methods of long-distance control: vessels, navigation and the Portuguese route to India' in Law, J. (ed.) *Power, Action and Belief: A New Sociology of Knowledge?*, London, Boston and Henley: Routledge and Kegan Paul, 234–63.
Law, J. (1986b) 'On power and its tactics: a view from the sociology of science', *Sociological Review*, 34, 1–38.
Law, J. (1999) 'After ANT: complexity, naming and topology' in Law, J. and Hassard, J. (eds) *Actor Network Theory and After*, Oxford: Blackwell, 1–14.
Law, J. and Hassard, J. (1999) (eds) *Actor Network Theory and After*, Oxford: Blackwell.
Lefebvre, H. (1968) *Dialectical Materialism* (trans. J. Sturrock), London: Jonathan Cape.
Lefebvre, H. (1991a) *The Production of Space* (trans. D. Nicholson Smith), Oxford: Blackwell.
Lefebvre, H. (1991b) *Critique of Everyday Life, Vol. I* (trans. J. Moore), London: Verso.
Lefebvre, H. (1996) *Writings on Cities* (trans. and eds E. Kofman and E. Lebas), Oxford: Blackwell.
Lipovetsky, G. (1994) *The Empire of Fashion: Dressing Modern Democracy* (trans. C. Porter), Princeton, NJ: Princeton University Press.
Lukes, S. (1974) *Power: A Radical View*, London: Macmillan.
McNay, L. (1994) *Foucault: A Critical Introduction*, Cambridge: Polity.
Mann, M. (1984) 'The autonomous power of the state: its origins, mechanisms and results', *Archives Européennes de Sociologie* 25, 185–213.
Mann, M. (1986) *The Sources of Social Power, Vol. I: A History of Power from the Beginning to AD 1760*, Cambridge: Cambridge University Press.
Mann, M. (1993) *The Sources of Social Power, Vol. II: The Rise of Classes and Nation States, 1760–1914*, Cambridge: Cambridge University Press.
Mann, M. (2001) 'Globalization and September 11', *New Left Review* 12, 51–72.
Massey, D. (1993) 'Power-geometry and a progressive sense of place' in Bird, J., Curtis, B., Putnam, T., Robertson, G. and Tickner, L. (eds) *Mapping the Futures: Local Cultures, Global Change*, London: Routledge, 59–69.
Massey, D. (1994) 'A global sense of place' in Massey, D. *Space, Place and Gender*, Oxford: Polity, 146–56.
Massey, D. (1999) *Power-Geometries and the Politics of Space-Time*, Hettner-Lecture, Department of Geography, University of Heidelberg.
Michie, R. C. (1992) *The City of London: Continuity and Change, 1850–1950*, London: Macmillan.
Miller, P. and Rose, N. (1990) 'Governing economic life', *Economy and Society* 19(1), 1–31.
Mills, C. W. (1956) *The Power Elite*, London, Oxford and New York: Oxford University Press.

Mitchell, T. (1988) *Colonizing Egypt*, Cambridge: Cambridge University Press.
Moretti, F. (2001) 'Planet Hollywood', *New Left Review* 9, 90–101.
Morley, D. and Robins, K. (1995) *Spaces of Identity: Global Media, Electronic Landscapes and Cultural Boundaries*, London and New York: Routledge.
Muir, R. (1997) *Political Geography: A New Introduction*, Basingstoke: Macmillian.
Murdoch, J. (1997) 'Towards a geography of heterogeneous associations', in *Progress in Human Geography* 21(3), 321–37.
Murdoch, J. (1998) 'The spaces of actor-network theory', *Geoforum* 29(4), 357–74.
Negri, A. (1991) *The Savage Anomaly: The Power of Spinoza's Metaphysics and Politics* (trans. M. Hardt), Minneapolis and Oxford: University of Minnesota Press.
Newman, D. (1999) (ed.) *Boundaries, Territory and Postmodernity*, London and Portland: Frank Cass.
O'Malley, P. (2000) 'Uncertain subjects: risks, liberalism and contract', *Economy and Society* 29(4), 460–84.
O'Malley, P., Weir, L. and Shearing, C. (1997) 'Governmentality, criticism, politics', *Economy and Society* 26(4), 501–17.
Ogborn, M. (1998) *Spaces of Modernity: London's Geographies, 1680–1780*, London and New York: Guilford Press.
Olds, K. (1995) 'Globalization and the production of new urban spaces: Pacific Rim megaprojects in the late twentieth century', *Environment and Planning A* 27, 1713–44.
Ong, A. and Nonini, D. (1997) (eds) *Ungrounded Empires: The Cultural Politics of Modern Chinese Transnationalisms*, New York: Routledge.
Osborne, T. (1998) *Aspects of Enlightenment: Social Theory and the Ethics of Truth*, London: UCL Press.
Osborne, T. and Rose, N. (1999) 'Governing cities: notes on the spatialisation of virtue', *Environment and Planning D: Society and Space* 17(6), 737–60.
Ó Tuathail, G. (1994) '(Dis)placing geopolitics: writing on the maps of global politics', *Environment and Planning D: Society and Space* 12(5), 525–46.
Ó Tuathail, G. (1996) *Critical Geopolitics: The Politics of Writing Global Space*, London: Routledge.
Ó Tuathail, G., Dalby, S. and Routledge, P. (1998) (eds) *The Geopolitics Reader*, London: Routledge.
Painter, J. (1995) *Politics, Geography and 'Political Geography'*, London: Arnold.
Painter, J. (2000) 'Pierre Bourdieu' in Crang, M. and Thrift, N. (eds) *Thinking Space*, London and New York: Routledge, 239–59.
Parsons, T. (1957) 'The distribution of power in American society', *World Politics* 10, 123–43.
Parsons, T. (1963) 'On the concept of political power', *Proceedings of the American Philosophical Society* 107, 232–62.
Peters, R. S. (1967) 'Authority' in Quinton, A. (ed.) *Political Philosophy*, Oxford: Oxford University Press, 83–96.
Peters, T. and Waterman, R. H. (1982) *In Search of Excellence*, New York: Harper and Row.

Philo, C. (1992) 'Foucault's geography', *Environment and Planning D: Society and Space* 10(2), 137–62.
Pile, S. (1997) 'Introduction: opposition, political identities and spaces of resistance' in Pile, S. and Keith, M. (eds) *Geographies of Resistance*, London: Routledge, 1–32.
Redding, S. G. (1990) *The Spirit of Chinese Capitalism*, New York: de Gruyter.
Rée, J. (2000) *I See a Voice: A Philosophical History*, London: Flamingo.
Richie, A. (1999) *Faust's Metropolis: A History of Berlin*, London: HarperCollins.
Robinson, J. (1996) *The Power of Apartheid: State, Power and Space in South African Cities*, Oxford: Butterworth Heinemann.
Rose, N. (1992) 'Governing the enterprising self' in Heelas, P. and Morris, P. (eds) *The Values of the Enterprise Culture: The Moral Debate*, London: Routledge, 141–64.
Rose, N. (1993) 'Government, authority and expertise in advanced liberalism', *Economy and Society* 22(3), 283–99.
Rose, N. (1994) 'Expertise and the government of conduct', *Studies in Law, Politics and Society* 14, 359–97.
Rose, N. (1996a) 'Authority and the genealogy of subjectivity' in Heelas, P., Lash, S. and Morris, P. (eds) *Detraditionalization*, Oxford: Blackwell, 294–327.
Rose, N. (1996b) 'Governing "advanced" liberal democracies' in Barry, A., Osborne, T. and Rose, N. (eds) *Foucault and Political Reason*, London: UCL Press, 37–64.
Rose, N. (1999) *Powers of Freedom: Reframing Political Thought*, Cambridge: Cambridge University Press.
Rose, N. and Miller, P. (1992) 'Political power beyond the state: problematics of government', *British Journal of Sociology* 43(2), 173–205.
Rosenau, J. N. (1997) *Along the Domestic–Foreign Frontier: Exploring Governance in a Turbulent World*, Cambridge: Cambridge University Press.
Said, E. W. (1978) *Orientalism*, New York: Pantheon.
Said, E. W. (1984) *The World, the Text and the Critic*, London and Boston: Faber and Faber.
Said, E. W. (1994) *Culture and Imperialism*, London: Vintage.
Sassen, S. (1991) *The Global City: New York, London, Tokyo*, Princeton, NJ: Princeton University Press.
Sassen, S. (1994) *Cities in a World Economy*, Thousand Oaks, CA: Pine Forge/Sage.
Sassen, S. (1995) 'On concentration and centrality in the global city' in Knox, P. L. and Taylor, P. J. (eds) *World Cities in a World System*, Cambridge: Cambridge University Press, 63–75.
Sassen, S. (1996) *Losing Control: Sovereignty in an Age of Globalization*, New York: Columbia University Press.
Sassen, S. (1998) *Globalization and its Discontents*, New York: New Press.
Sassen, S. (1999) 'Digital networks and power' in Featherstone, M. and Lash, S. (eds) *Spaces of Culture: City, Nature, World*, London and Thousand Oaks, CA: Sage, 49–63.
Sassen, S. (2000) 'Territory and territoriality in the global economy', *International Sociology* 15(2), 372–93.

Sayer, A. (1992) *Method in Social Science: A Realist Approach*, 2nd edition, London and New York: Routledge.
Sayer, A. (2000) *Realism and Social Science*, London and Thousand Oaks, CA: Sage.
Scott, A. and Street, J. (2000) 'The use of popular culture and new media in parties and social movements', *Information, Communication and Society* 3(2), 215–40.
Scott, D. (1995) 'Colonial governmentality', *Social Text* 5(3), 191–220.
Sennett, R. (1980) *Authority*, New York and London: W. W. Norton.
Sennett, R. (1990) *The Conscience of the Eye*, London and Boston: Faber and Faber.
Serres, M. and Latour, B. (1995) *Conversations on Science, Culture and Time* (trans. R. Lapidus), Michigan: University of Michigan Press.
Sharp, J. P., Routledge, P., Philo, C. and Paddison, R. (2000) *Entanglements of Power: Geographies of Domination/Resistance*, London and New York: Routledge.
Sheridan, A. (1980) *Michel Foucault: The Will to Truth*, London and New York: Tavistock.
Shields, R. (1992) 'A truant proximity: presence and absence in the space of modernity', *Environment and Planning D: Society and Space* 10(2), 181–98.
Shields, R. (1999) *Lefebvre, Love and Struggle: Spatial Dialectics*, London and New York: Routledge.
Simmel, G. (1997) 'Bridge and door', in Frisby, D. and Featherstone, M. (eds), *Simmel on Culture: Selected Writings*, London and Thousand Oaks, CA: Sage, 170–4.
Spivak, G. C. (1996) 'More on power/knowledge' in Landry, D. and Maclean, G. (eds) *The Spivak Reader: Selected Works of Gayatri Chakravorty Spivak*, New York and London: Routledge, 141–74.
St Clair, J. (1999) 'Seattle diary', *New Left Review* 238, 81–96.
Swyngedouw, E. (1997) 'Neither global nor local: "glocalisation" and the politics of scale' in Cox, K. (ed.) *Spaces of Globalization: Reasserting the Power of the Local*, London and New York: Guilford Press, 137–66.
Swyngedouw, E. (2000) 'Authoritarian governance, power, and the politics of rescaling', *Environment and Planning D: Society and Space* 18(1), 63–76.
Taylor, C. (1985) 'Foucault on freedom and truth' in Taylor, C. *Philosophy and the Human Sciences, Philosophical Papers, Vol. 2*, Cambridge: Cambridge University Press, 153–84.
Thompson, J. (1995) *The Media and Modernity*, Cambridge: Polity.
Thrift, N. (1994) 'On the social and cultural determinants of international financial centres: the case of the City of London' in Corbridge, S., Martin, R. and Thrift, N. (eds) *Money, Power and Space*, Oxford: Blackwell, 325–55.
Thrift, N. (1999) 'Steps to an ecology of place' in Massey, D., Allen, J. and Sarre, P. (eds) *Human Geography Today*, Cambridge: Polity, 295–322.
Tomlinson, J. (1999) *Globalization and Culture*, Cambridge: Polity.
Virillio, P. (1986) *Speed and Politics*, New York: Semiotext(e).
Virillio, P. (1991) *The Lost Dimension* (trans. D. Moskenberg), New York: Semiotext(e).
Virillio, P. (1997) *Open Sky* (trans. J. Rose), London and New York: Verso.
Virillio, P. (2000) *Polar Inertia* (trans. P. Camiller), London and Thousand Oaks, CA: Sage.

Walter, E. V. (1973) 'Simmel's sociology of power: the architecture of politics' in Wolff, K. H. (ed.) *Georg Simmel: Essays on Sociology, Philosophy and Aesthetics*, New York: Harper and Row, 139–66.

Weber, M. (1978) *Economy and Society*, vols 1 and 2 (eds G. Roth and C. Wittich), New York: Bedminster Press.

Wolff, K. H. (1950) *The Sociology of Georg Simmel*, New York: Free Press.

Young, I. M. (1990) *Justice and the Politics of Difference*, Princeton, NJ: Princeton University Press.

Index

actor network theory 86, 142, 147
administration *see* bureaucracy
advertising
   influence of 21, 30–1
   Monsanto's use of 121–2
   working of 98, 102, 103–4, 146
Agnew, John 33–4
Aksoy, Asu 109
Amnesty International 55
architecture *see* buildings
Arendt, Hannah
   on authority 29, 58, 114, 126
   on collective power 52–9, 106, 123–6, 173, 184
   on empowerment 39, 54–5
   on persuasion 58, 125
   on power 2, 4, 20, 30, 100, 111–12
   on power relations 102
   on violence 57, 184
armed forces
   and control 71
   power of 49–50
Aron, Raymond 107–8
arrangement, of power
   diagrams of power 69–75
   distribution of power 24–5, 32–6, 46–7, 59–64, 83–90
   proximity and reach 129–58, 183, 192

   in social spaces 159–88
Art and Revolution 184
associational power 20, 51–9, 106–7, 123–7, 183–7, 196–7
   see also Arendt, Hannah, on collective power; collective action; empowerment
asymmetry, and power 26–7
authority
   Arendt on 29, 58, 114, 126
   and bureaucracy 118–19
   charismatic 119–20, 149
   and City of London 181
   and domination 6, 29–30, 101
   exercise of 118–20
   governments' use of 122–3, 139–51
   Monsanto's use of 121–2
   nature of 126
   people's trust in 122–3, 142–3, 147–8, 156, 168–70, 183
   and politics 120–1
   and power 6
   reach of 130, 148–51, 156, 183, 192
   relationship to practices 117
   and Rockefeller centre 114
   and seduction 121, 122
   and social space 11, 168–9, 171
   and spatiality 3, 10
   traditional 119, 121–2, 169, 181

INDEX

Bank of England 171
banking *see* financial institutions
Barnes, Barry 118, 134
Barnett, Clive 148
Beaverstock, Jonathan 152
Benhabib, Seyla 54, 56, 57, 124
Bentham, Jeremy 70
Berlin, seductive qualities of 176–8, 181–2
Bhaskar, Roy 22, 23
biotechnology industry, influence of 121–3, 124–5
Black Bloc 184
Boden, Deirdre 137, 148
borders and boundaries
　Lefebvre on 163
　spatial enclosures 172–4
Bourdieu, Pierre 38, 41, 44
branding, corporate 176
broadcasting 138
Bromley, Simon 184
buildings
　as seducers 176–8, 182
　and sense of power 111, 113–15, 169, 194
　as surrounders of spaces 163
bureaucracy
　and administration from afar 132–4, 141
　Arendt on 57–8
　and authority 118–19
　bureaucratization of space 167
　and domination 32, 57–8
　working of 32–3

Caldeira, Teresa 172–3
Callon, Michel 86, 113, 133, 142, 147
campaigns, popular 55–7, 58, 124–5, 184–7
capacity, power as a 16–25
Carnival Against Capital 185
Castells, Manuel 7, 8, 60–3, 64, 112–13, 115
CCTV *see* closed-circuit television cameras
centred power 113–16

distribution of 24–5, 32–6
multi-centred power 35–6
charisma 119–20, 149
China
　overseas Chinese networks 62–3
　Tiananmen Square incident (1989) 57
Christian Aid 125
cinema *see* film industry
cities
　gated communities 172–3, 174, 195
　as places of seduction 176–8
　and power 61–2, 74–5, 111–13
citizenship, and authority 144–5, 146, 147–9
City of London
　influence of 28, 111, 155–6
　and power 181
　protests against 185
　as social space 161–2, 163–6, 168–72
closed-circuit television cameras (CCTV) 138, 175
Cochrane, Allan 176
codes, Lefebvre on 162–6
coercion
　and advertising 31
　and government 143
　nature of 31, 101
　reach of 139
　relationship to practices 117
　and seduction 122
　and spatiality 10
　in the workplace 145, 149–50
collective action 38–64, 123–7
　see also associational power; communities; networks
colonization
　as analogy for space domination 167–8
　operation of 85, 131–2, 133
communications
　corporate 157
　face-to-face vs electronic 137, 148
　and power 135–9, 151–2
　see also internet; networks, electronic communities

capacity for domination 18
gated communities 172–3, 174, 195
and power 5–6, 10–12
see also associational power;
  companies; institutions; networks;
  relational ties
companies
  control of workforce 144–6, 149–50
  corporate branding 176
  corporate communications 157
  corporate culture 144–6, 149–50
  and proximity 192–3
  see also City of London; institutions
companies, multinational
  judging power of 17–18
  nature of power of 23–5, 33, 34, 47
  reach of 152–7
  and the state 8
  see also globalization
competition, and power 117–18
confession, Foucault on 76–7, 82
The Conscience of the Eye (Sennett) 114
consent, and power 20–1
control of behaviour 65–91, 132–5,
  139–51, 167
Cooper, Davina 140
corporations *see* companies
Crang, Mike 181
culture, corporate 144–6, 149–50
culture, influencing 30–1, 33, 50

Dahl, Robert 17
Daimler Chrysler corporation 176, 182
Dean, Mitchell 79, 86
Debis quarter, Berlin 176, 182
Debord, Guy 167
debt relief campaigns 55
decentred power 134–5
  power as immanent force 65–91
  power in spaces 159–88
  see also associational power
delegation 24–5, 32–3, 118, 134
Deleuze, Gilles
  diagrams of power 69–70, 73–4, 90
  on institutional power 82–3
  on power and space and time 8–9, 191

on power's nature 7, 66
on power's volatility 84–5, 134
Disch, Lisa Jane 54, 124
discipline
  Hindess on 77
  institutional 69–75
  and power 24–5, 32–3
Discipline and Punish (Foucault) 70–1
Discourses of Power (Hindess) 18–21
distraction 176, 177
distribution of power 24–5, 32–6, 46–7,
  59–64, 83–90
domination
  and authority 6, 29–30, 101
  and bureaucracy 32, 57–8
  cultural 30–1
  and government 100
  and institutions 69–75
  as means of influencing others 78
  mediation of 130
  nature of 166
  potential capacity for 16–18, 19–20
  and power 19–20, 25–6, 30, 190
  reach of 183, 192
  relationship to practices 117
  and relationships 26–7
  and Rockefeller centre 114
  and seduction 30–1, 101, 102,
    122
  of social space 11, 162–75
  and spatiality 10
  sustainability of 29–30
  Weber on 26, 27–32, 87, 101
Driver, Felix 72
du Gay, P. 71, 149

economic institutions, influence
  of 48–50
Elden, Stuart 71
elites, origin of power 61–2
Empire (Hardt and Negri) 86–9
empowerment
  and collective action 54–5, 196
  and mobilization of resources 38–9
  and understanding of power 12, 39
enclosures, spatial 163, 172–4

entertainment 176, 177
  see also film industry
environmental groups, power of 55–7, 58, 124–5, 184–6, 187
experts, Rose on 139–53
  see also authority

family life, Foucault on 80–1
feminism, on power 55
festivals, Lefebvre on 167
film industry, influence of 30–1, 33, 109, 111
financial institutions
  influence of 28, 29–30, 154–7, 180
  see also City of London
Financial Times 156
food, genetically modified 121–3, 124–5
force, threat of *see* coercion
Foucault, Michel
  and Hindess 19
  influence on Rose 150
  on *pouvoir* 108–9
  on power and institutions 7
  on power and self-government 41, 65–85, 90, 98–100
  on power and space and time 8, 191
  on power as enabler 124
  on power relations 122
  on provenance of power 1
  on violence 100
Friends of the Earth 55–6
fundamentalism, Said on 36

gated communities 172–3, 174, 195
Gates, Bill 106
genetically modified food 121–3, 124–5
Geographical Imaginations (Gregory) 167
geographical information systems 138
Giddens, Anthony
  on authority figures 58, 126, 142–3
  on distanciated power 7, 44
  Hindess on 20
  on networks and power 59
  on Parsons 42, 43, 52
  on power and social relations 44, 55
  and power as capacity 21
  and power as facilitative 39
  on proximity 135
  on resources and power 5, 43–7, 96, 114–15
globalization
  and networks 86–90
  and power 8, 35–6, 51, 52
  and social interaction 46
  see also companies, multinational
Gordon, Colin 77, 79, 85, 90–1
government and governments
  control of citizens 74, 75–91
  government by domination 100
  government by information 132–4
  government by rules 68
  government from afar 139–51, 167
  government sites 113–16
  origin of power 57
  see also self-government; states
Greenpeace 55–6, 57
Gregory, Derek 68–9, 160, 167–8
Guattari, Felix 7, 66, 73, 74, 82–3, 134

Habermas, Jürgen 53, 125
Hacking, Ian 77
Hardt, Michael 86–9, 194
Harré, Rom 22
Hartsock, Nancy 54
Harvey, David 34
Haus Huth, Berlin 182
Hegel, Georg 161
Heidegger, Martin 124
Hindess, Barry
  on discipline 77
  on domination and government 100
  on power as capacity 16, 18–21, 25, 27, 105, 196
history, as a resource 181–2
Hollywood, influence of 30–1, 33, 109–10, 111
Howe, Nicholas 176
Howell, P. 53

INDEX

The Human Condition (Arendt) 53, 124
human rights activists 55
Huyssen, Andreas 176

ideology, influencing 50
imagery, as control mechanism 67
images, out-of-scale 36–7
IMF *see* International Monetary Fund
inducement
  Foucault on 173
  and globalization 155–6
  and government 123, 132–4, 143, 144
  and money 41
  nature of 101
  relationship to practices 117
  Rose on 147
  in the workplace 145, 149
information, and power 45–6, 60–2, 112–13
institutions
  authority within 118–19
  and control 69–73, 82–3
  dispersement of responsibilities 98
  economic 48–50
  financial 28, 29–30
  military 49–50, 71
  and power 7, 9, 15–29, 32–7, 47–52, 66
  sites of 113–16
  see also City of London; companies; companies, multinational; networks
instrumental power 117–23
interaction
  face-to-face vs electronic 137
  and power 38–64
  and states 45–6
  see also relational ties
International Campaign to Ban Landmines 55
International Monetary Fund (IMF) 8, 26
internet
  and association 55, 60–2, 184
  and control 137–8
  see also networks, electronic
Isaac, Jeffrey 21–8, 32

Jessop, Bob 34
Jonas, Andrew 176
Jubilee 2000 debt relief campaign 55

Kern, Stephen 135
Klein, Naomi 88–9, 195
Kracauer, Siegfried 177

landmines, campaigns against 55
Latour, Bruno
  influence on Rose 86, 140, 145, 147, 150–1, 157, 158
  on power as thing 15–16
  on power's movement 97, 113
  on power's reach 129–34, 141–2, 158
Lefebvre, Henri 10–11, 160–72, 185–7
Lipovetsky, Gilles 30–1, 50, 103, 120
London, City of *see* City of London
Lukes, Steven 7

Machiavelli, Niccoló 5
McNay, Lois 76, 77, 78
Madden, Ed 22
manipulation
  and advertising 31
  and film industry 110
  and government 143, 144
  nature of 31, 148–9, 150
  reach of 139, 150, 192
  and seduction 102
  and social space 11, 170, 171
  and spatiality 10
  in the workplace 145–6, 150
Mann, Michael
  Hindess on 20
  on power and networks 7, 8, 43–4, 46–52, 59, 63, 96–7
  and power as capacity 21
  on power's reach 139
  on state resources 115
Marx, Karl 21, 161
mass media 138

Massey, Doreen 174
media, broadcast and mass 138
mediation
  of power 97–8, 130–1
  of power modes 102, 122
membership
  of gated communities 172–3
  Lefebvre on 163–4
Mexico 154
Microsoft 106
middle classes
  and capacity for power 18, 33
  origin of power 40
military institutions *see* armed forces
Miller, P. 68, 86
Mills, C. Wright 7, 40–1
Mitchell, Timothy 168
modes of power *see* authority; coercion; domination; inducement; manipulation; negotiation; persuasion; seduction
Molotch, Harvey 137, 148
money, as analogy for power 41–2
Monsanto 121–3
monumental buildings, and sense of power 111, 113–15, 169, 194
moral governance 67–8, 74
Mosca, Gaetano 7
Mosley, Oswald 119
motivation, of workforce 144–6, 148, 149–50
multinational companies *see* companies, multinational
multi-tier governance 34–6
mutual action, and power 53

nation states *see* states
negotiation
  Arendt on 58
  nature of 125–6
  Rose on 147
  in the workplace 146, 150
Negri, Antonio 86–9, 194
networks
  Castells on 7, 8, 60–3, 64, 112–13, 115
  and globalization 86–90
  and government from afar 132–4
  and power 7, 8, 38–64
networks, electronic
  effect of 131
  and globalization 154–7
New York, architecture of 114
Newman, David 34
non-governmental organizations (NGOs) 8

O'Malley, Pat 142
organizations *see* companies; institutions; non-governmental organizations
Orientalism (Said) 68–9
Osborne, Thomas 74–5, 119
Oxfam 125

Pareto, Vilfredo 7
Parsons, Talcott 5, 20, 39, 40–3, 51–2, 53
past, as a resource 181–2
pastorship, Foucault on 81–2
persuasion
  Arendt on 58, 125
  and environmental groups 56
  and globalization 156
  and government 144, 148
  nature of 50, 125–6
  Rose on 147
  and social space 171
Peters, Tom 145
phenomenology, of power 159–88
Philo, Chris 72
Pile, S. 78
place
  and authority 11, 168–9, 171
  and domination 11, 162–75
  and manipulation 11, 170, 171
  and persuasion 171
  power's reach 129–58, 183, 192
  presence and power 180–3
  and seduction 170–1
  see also proximity
Poland, democracy movement 187

politics
    people power 54–7, 58, 124–5,
      184–7
    personality politics 119–21
    see also states
potentia, Negri on 89
potestas, Negri on 89
Potsdamer Platz, Berlin 176–8, 181–2
pouvoir 107, 108–9
power
    associational 20, 51–9, 106–7,
      123–7, 183–7, 196–7
    authoritative 48–9
    as capacity 16–25
    definition of 2, 4–6, 8
    diagrams of 69–75
    diffused 48–50
    distribution of 24–5, 32–6, 46–7,
      59–64, 83–90
    etymology 105–6, 107–8
    guises of 95–128
    as immanent force 65–91
    instrumental 117–23
    mediated nature of 97–8
    mediated nature of modal
      qualities 102
    as medium 39–44, 59
    spatial vocabulary of, definition 13
    as technique 65–8
    as thing 8, 9, 15–37, 110–13
Powers of Freedom (Rose) 139–53
presence, and power 180–3
prisons, discipline in 70–3
The Production of Space (Lefebvre)
    160–72
proximity
    and authority 148–53
    and companies 192–3
    and power 129–58
    social vs spatial 135–9
public space, Arendt on 54–5, 56, 184
puissance 107–8

Rainbow Warrior 57
Rainforest Action Network 184
Rée, Jonathan 169

relational ties
    collaborative 5–6
    and domination 26–7
    instrumental 5–6
    interaction and power 38–64
    and power 21–3, 26, 44, 55, 59
    and social space 171–2
resistance 78, 195–6
    see also campaigns, popular
resources
    allocative 44–5, 107
    authoritative 44–5, 107
    and empowerment 38–9
    and power 5, 44–51, 59, 63, 96,
      105–16, 194
Reuters 156
rhythms, of power 160–72
Richie, Alexandra 176
The Rise of the Network Society
    (Castells) 60–3, 64
Robins, Kevin 109
Rockefeller centre, New York 114
Rose, Nikolas 68, 74–5, 79, 86,
    139–53, 157
    Latour's influence 86, 140, 145, 147,
      150–1, 157, 158
Rosenau, James 34
Royal Society for the Protection of
    Birds 125
rules
    and control 67–8
    and membership 172–3
    and power 24–5, 32–3, 116–17

Said, Edward 36, 68
Sassen, Saskia 112, 115, 152–7
Scott, Alan 185
Scott, David 85, 88
Seattle, protests against WTO
    (1999) 184–5
seduction
    and authority 121, 122
    and coercion 122
    and domination 30–1, 101, 102, 122
    and film industry 110
    and globalization 156

seduction (*cont'd*)
  and government 143, 144, 148
  and manipulation 102
  mediation into other power modes 102, 122
  nature of 30–1, 50, 148, 178, 190
  in open spaces 175–8, 182
  and politics 120–1
  reach of 103, 104, 130, 139, 150, 183, 192
  Rose on 147
  and social space 11, 170–1
  and spatiality 10
  in the workplace 146, 149
self-definition 38
self-government
  Foucault on 41, 65–8, 75–91
  Rose on 144–53
Sennett, Richard 114
senses, and domination of space 162, 164, 169
Serres, Michel 158
service work, nature of 165–6, 171
sexuality, control of 67, 74
Sharp, Joanne 78
Sheridan, Alan 99, 100, 101
Shields, Rob 135, 160, 167
Simmel, Georg 163, 186
Situationists 167, 185
smell, and domination of space 162, 164
Smith, Richard 152
social groups *see* communities; networks
social relations *see* relational ties
social welfare 141
Soil Association 125
Solidarity (Polish trade union movement) 187
Sony Centre, Berlin 176–8, 182
sound, and domination of space 162, 164, 169
space
  and authority 11, 168–9, 171
  counter-spaces 185–7
  domination of 162–75
  induced and produced 166
  manipulation of 11, 170, 171
  open spaces 174–8
  and persuasion 171
  and power 10–12
  public, Arendt on 54–5, 56, 184
  representational 160–1
  representations of 160–1
  and seduction 11, 170–1
  spatial enclosure 163, 172–4
spatiality
  and effect of different power modes 102–5
  and institutional control 70–1
  and power 3, 9–12, 159–97
  of power 129–58
  spatial vocabulary of power, definition 13
spectacles, Lefebvre on 167
Spinoza, Benedict de 89
Spivak, Gayatri Chakravorty 109
states
  Castells on 115
  and globalization 8
  influence of financial institutions 28
  and multinational companies 154
  origin of power 57
  and power 33–6, 48–9
  resources 115
  and social interaction 45–6
  see also government and governments
Street, John 185
'The subject and power' (Foucault) 75–6
supervision, and power 24–5, 32–3
surveillance, electronic 138, 175

Taylor, Charles 67, 80
Taylor, Peter 152
technology
  and power 135–9, 151–2, 175
  see also internet; networks, electronic
telecommunications *see* communications
terrorism, Said on 36
Thompson, John 137
Thrift, Nigel 155–6, 161

Tomlinson, John 137
topology, of power 65–91, 191–3
　importance of 192
　place and power 162–75
　power's reach 129–58, 183, 192
　presence and power 180–3
　see also proximity; space; spatiality
totalitarianism, Arendt on 57
tradition, and authority 119, 121–2, 169, 181
tyranny, Arendt on 57

United Kingdom government
　and Monsanto 122–3
　policies 145
United States, personality politics 120

violence
　Arendt on 57, 184
　Foucault on 100
　at Seattle 184
Virilio, Paul 157

Waterman, R. H. 145
Weber, Max
　on authority 58, 149
　on bureaucracy 57, 118–20
　on domination 26, 27–32, 87, 101
　on economic liberalism 81
　on global economy 153
　and Hindess 19
　on instrumental power 117–18
　on power 2, 4, 7, 100
　on power as capacity 21
　on power as thing 8, 9
work
　power in workplace 144–6, 148, 149–50
　service work, nature of 165–6, 171
World Bank 8
World Trade Organization (WTO),
　Seattle protests (1999) 184–5

Young, Iris Marion 54